AF841569

A JAPANESE APPROACH TO POLITICAL ECONOMY

Also by Robert Albritton

A JAPANESE APPROACH TO STAGES OF CAPITALIST DEVELOPMENT
A JAPANESE RECONSTRUCTION OF MARXIST THEORY

Also by Thomas T. Sekine

THE DIALECTIC OF CAPITAL, 2 vols

A Japanese Approach to Political Economy

Unoist Variations

Edited by

Robert Albritton
Associate Professor of Political Science
York University, North York
Canada

and

Thomas T. Sekine
Professor of Economics and Social and Political Thought
York University, North York
Canada

M

St. Martin's Press

© Robert Albritton and Thomas T. Sekine 1995

All rights reserved. No reproduction, copy or transmission of this publication may be made without written permission.

No paragraph of this publication may be reproduced, copied or transmitted save with written permission or in accordance with the provisions of the Copyright, Designs and Patents Act 1988, or under the terms of any licence permitting limited copying issued by the Copyright Licensing Agency, 90 Tottenham Court Road, London W1P 9HE.

Any person who does any unauthorised act in relation to this publication may be liable to criminal prosecution and civil claims for damages.

First published in Great Britain 1995 by
MACMILLAN PRESS LTD
Houndmills, Basingstoke, Hampshire RG21 2XS
and London
Companies and representatives
throughout the world

A catalogue record for this book is available
from the British Library.

ISBN 0–333–62796–2

10 9 8 7 6 5 4 3 2 1
04 03 02 01 00 99 98 97 96 95

Printed and bound in Great Britain by
Antony Rowe Ltd
Chippenham, Wiltshire

First published in the United States of America 1995 by
Scholarly and Reference Division,
ST. MARTIN'S PRESS, INC.,
175 Fifth Avenue,
New York, N.Y. 10010

ISBN 0–312–12435–X

Library of Congress Cataloging-in-Publication Data
A Japanese approach to political economy : Unoist variations / edited by Robert Albritton and Thomas T. Sekine.
p. cm.
Includes bibliographical references and index.
ISBN 0–312–12435–X
1. Economics—Japan. 2. Economics—Philosophy. 3. Capitalism.
I. Albritton, Robert, 1941– . II. Sekine, Thomas T. (Thomas Tomohiko), 1933– .
I. Title.
HB126.J2J37 1995
330'.0952—dc20 94–30589
CIP

Contents

Notes on the Contributors

Robert Albritton, Associate Professor, Political Science, York University, Toronto.

John R. Bell, Professor, School of Liberal Studies, Seneca College, Toronto.

Colin Duncan, Lecturer, History, Queen's University, Kingston, Ontario.

Tomiichi Hoshino, Professor, Economics, Toyama University, Japan.

Sadao Ishibashi, Professor, Economics, Kyushu Sangyo University, Japan.

Makoto Maruyama, Associate Professor, Social Sciences, University of Tokyo, Japan.

Thomas T. Sekine, Professor, Economics and Social and Political Thought, York University, Toronto.

Randall Terada, Graduate Student, Political Science, York University, Toronto.

1 Introduction

Robert Albritton and John R. Bell

What all of the authors who have contributed essays to this anthology have in common is that they have studied with or been influenced by the work of Thomas Sekine to various degrees. Sekine studied with one of the most prominent and influential Marxian political economists in post-World War II Japan: Kozo Uno (1897–1977). He then left Japan and studied in Canada and England, finally settling in to a teaching career at York University, Toronto, where he has taught since 1968. Although the once large and significant Japanese Uno School is today largely in a state of decline within Japan, this book attests to the fact that at least one branch of it is very much alive and vital. Ironically, given the rootedness of Uno Theory in Japan, it is arguably the Canadian branch that today is doing the most to develop the legacy of Uno. And this is largely due to Sekine's theoretical contributions.

There are few if any thinkers who have contributed more since Marx himself to the clarification of the theory of capital than Uno. But the full significance of Uno's work has not always been appreciated even by Unoists themselves. In Japan Uno is respected, but a great deal of time is wasted in disagreeing with particular points in his theory rather than improving the clarity and rigor of his theory as a whole. In our view it is Sekine who has done the most to bring out the exciting potentialities of Uno's work. He has done this both by making explicit the dialectical logic that was implicit in Uno's *Principles of Political Economy* (1980), and by providing us the most clear and rigorous restatement and commentary on Uno's *Principles* to date in his *Dialectic of Capital* (2 vols) (Sekine, 1984, 1986). In the area of epistemology Sekine's work not only grasps the specific uniqueness of capital as an object of knowledge, but also as a result of this breakthrough helps us to appreciate the potential power of Marxian political economy. This power comes from being clear about the self-reifying and self-objectifying character of capital, a clarity which indicates the possibility and desirability of constructing a rigorous theory of capital's logic, while at the same time problematizing the interrelations between this logic and socio-historical life considered more concretely. And by this clarification of the object of knowledge of

Marxian political economy, Sekine's work also helps those who are engaged in an exploration of the links which can be established between Marxian political economy and social science as a whole.

Few Western Marxists have seriously studied the work of Uno and Sekine. In some ways Kozo Uno is to Japanese Marxism as Louis Althusser is to Western Marxism. Both were very influential in the 60s and 70s and have continuing legacies of influence. Both emphasized that Marx's *Capital* was in some sense the founding work of a new science, and hence of central and crucial importance within the corpus of Marx and Engels' writings. Both had conceptions of levels of analysis within Marxist theory. Both were criticized for excessive formalism and for creating a gulf between theory and practice. The analogy may, however, obscure more than it illuminates if we stop with these similarities and ignore some crucial differences stemming perhaps most fundamentally from the fact that Althusser was primarily a philosopher and Uno an economist.

As a philosopher Althusser was able to explore the "epistemological breaks" separating Marx's *Capital* from previous works of economic theory and more broadly Marx's epistemology from those "problematics" that it broke with, but he was not able to make *substantive* contributions to the Marxian economic theory of capitalism or to political economy generally. But getting the substantive theory right is a prerequisite to being clear about epistemological questions. What Uno did was nothing less than to reconstruct Marx's *Capital* as a whole into a rigorous theory of capital's basic operating principles. This not only established a framework for getting the theory of capital "right", but also made it clear that the relation between the economic and the non-economic and between abstract economic theory and economic history is a highly problematic one requiring distinct levels of theory.

Lacking a rigorous theory of capital's logic, Althusser tended to base his epistemological conclusions on *Capital*, which, as a "founding" work of a research program, is still quite embryonic and incomplete as an economic theory. Althusser is able to show that "surplus value" represents an important epistemological break, but he is not able to enlarge, refine, and ultimately defend the economic theory in which "surplus value" plays a central role. As a result, Althusser's epistemology is not so much grounded in a well formulated substantive theory of capital as it is in distancing "mature" Marxist epistemological positions from epistemological "problematics" that Marx is supposed to have broken with. The result is a highly negative epistemology. Thus Althusser's version of Marxian epistemology is anti-essentialist, anti-humanist, anti-historicist, anti-economistic, anti-reductionist, anti-empiricist, and anti-rationalist. But without a rigorous reformulation of Marx's economic theory that is scientifically defensible, Marx's *Capital* itself is open to the radically dissolving and undermining tendencies of an epistemology based on a series of "anti's". For this reason Althusserian Marxism cannot withstand the withering anti-essentialist attacks of postmodernism; indeed, however much Althusser himself was committed to Marxism, his philosophy might be seen as a major progenitor of the postmodernism that has done so much to dissolve Marxism. In our view because Althusser was unable to think rigorously through and where necessary reconstruct Marx's *Capital,* he was also unable to understand fully the form of scientificity peculiar to capital as an object of knowledge nor was he able to specify correctly the "epistemological breaks" peculiar to this scientificity. The result was a disasterous scientific pretentiousness, which set Western Marxism on a course that was almost certain to end in its deconstruction.

In contrast to Althusserian Marxism, Unoist political economy not only can withstand the onslaught of anti-essentialism, but, where constructive, can absorb its criticisms, and, at the same time build upon the tremendous power of Marx's *Capital.*[1] As an economist, Uno was able to make an enormous contribution to the substantive content of Marxian political economy, but his weakness was his inability to bring out the full epistemological implications of his work. For opening up this whole area we need to thank Sekine.

Marxian economic theory has been taken much more seriously in Japan than in North America. It has been estimated that at its height in the late 1960s perhaps as many as 50% of all professors of economics in Japanese Universities were primarily oriented towards Marxian economic theory (today the percentage is much smaller).[2] Uno influenced a whole generation of Marxian economists in the 1960s and 1970s.[3] By the early 1980s there were Unoist economists at many of the leading Japanese Universities. At the same time, the Uno School became internally complex as various sub-schools developed. Unoists tended to agree on the importance of theorizing Marx's law of value in the context of a purely capitalist society, but they disagreed on exactly how the law of value was to be formulated and on how this highly abstract economic theory was to relate to history. One of the major divisions, which had to do with this latter issue, was between the "world capitalism" school, which emphasized a closer connection between abstract economic theory and history than other levels of analysis approaches that emphasized the need for three distinct levels of theory. The work of the "world capital" school that is best known in the west is that of Makoto Itoh (1988). Whereas the work of the "levels of analysis" approach that is best known is that of Sekine (1984, 1986) and Albritton (1986, 1991). It is important to keep in mind that the essays in this volume are not representative of the Uno School as a whole, but only of one branch.[4] And even this one branch is internally complex, as the variety of appropriations of Uno's and Sekine's ideas that are represented in this volume demonstrate.

Uno's greatest contributions to Marxian political economy are, first, his reformulation of Marx's *Capital* as a logically rigorous theory of a purely capitalist society and, second, his idea that there should be three distinct levels of theory: the theory of basic principles (pure capitalism), a mid-range theory of stages, and historical/empirical analysis. Uno so thoroughly mastered Marx's economic thought that he had the confidence to reconstruct it as a whole to be more in accord with its ultimate aim, which was to theorize accurately capital's logic. Or to put it a little differently, by totally immersing himself in Marx's economic thought, Uno was better able than Marx to grasp fully the necessary logic of capital itself when it is not interferred with by extra-economic force or historical contingency. The result was an "objective" theory in the sense that he let the logic of capital emerge as a commodity-economic logic that governs the reproduction of a purely capitalist society (all production by commodified labour power).[5] Because capital cannot by itself theorize its own movement the theorist must, in a manner of speaking, "help it along", but when the theorist constructs an artificial environment where capital is not interfered with, the result is a theory in which, to some extent, the logic of capital reveals itself. Thus, while adopting the basic framework of *Capital*, Uno makes the theory much more logically coherent and more in accord with capital's own basic operating principles. This often requires certain changes in the sequencing of categories or in their modes of presentation, and, of course, most fundamentally, rigorous adherence to a totally commodified society in which the commodification of labour power and all other inputs and outputs of economic life can be managed by a commodity-economic logic without any reliance upon extra-economic force.

The advantage of this self-conscious surrender to capital's logic is that one can allow capital's logic to tell its own story at the level of pure theory, then step back and examine it from the outside at the level of history so as to see those aspects of social life which capital obscures or to criticize aspects of capital's impact on social life. The theory of pure capitalism, then, gives us knowledge that is limited by the fact that pure capitalism never exists in history, and yet this knowledge can play a crucial role in sorting out the role that capital actually does play in the impure and complex capitalisms that actually exist. By contrast neoclassical economists are not conscious of their surrender to capital's logic, and hence cannot avoid perpetrating what Polanyi (1944) has aptly called a "market mentality".

Uno devoted most of his attention to his theory of pure capitalism; hence his idea of levels of analysis was left relatively undeveloped. The idea of levels of analysis may seem fairly commonplace both within and outside Marxian political economy. What is different about Uno's conceptualization is the insistence that each level requires a completely distinct theory because the degree of necessity that operates at each level differs. Thus, for example, stage theory and historical analysis may inform one another, but, as separate theories with their own concepts, they cannot be simply deduced or induced from one another. Uno not only makes levels of analysis important as never before, but also his insistence on the separate kinds of theoretical logics operating at each level problematizes levels of analysis in a new way. While Uno did not go very far in specifying the relations between the levels, he does formulate helpful hints. For example, a major work entitled *Types of Economic Policy under Capitalism*, which has not yet been published in English, theorizes three stages of capitalism: mercantilism, liberalism, and imperialism. Where at the level of pure capitalism, economic life is governed by self-expanding value, at the more concrete level of stage theory, the logic of capital is translated into a more concrete institutional configuration of capital accumulation accompanied by stage-specific types of policy.

Uno's conception of stage theory may superficially appear to be largely an adaptation of Lenin's notion that imperialism is the highest *stage* of capitalism. But Uno's notion of stages is fundamentally different from Lenin's. The most important difference is that Uno considers stage theory to be a separate level of theory from the theory of capital's logic. Hence, for Uno the movement from one stage to the next cannot be a simple extension of capital's laws of motion. For example, Uno would argue that as long as the production and consumption of light use-values such as cotton prevailed there would be no tendency for the laws of motion of capital by themselves to produce monopoly or oligopoly. Thus, the stage of imperialism arose primarily as a result of the historically contingent development of heavy and expensive steel-making technologies which small competitive firms could not afford to implement. Thus, with the cooperation of political authorities the limited liability joint-stock company assumed a leading role as the form of capital best suited to operate these new industries. Thus, for Uno, at the level of stage theory, there is no inner logic or teleology that necessarily produces a succession of stages. It follows, that in order to understand the transition from one stage to the next, Uno posits not some iron law at a higher level of abstraction, but instead turns his attention to the analysis of economic history to understand the contingent development of a new complex of use-value production and accompanying technology.

Sekine's greater exposure to Western philosophy and neo-classical economic theory has enabled him to enlarge and refine Uno's theory significantly. In particular, as previously mentioned, Sekine makes explicit the dialectical logic implicit in Uno's theory of capital's basic operating principles. Indeed, Sekine has discovered such a close correspondence between Hegel's *Logic* (1969) and capital's logic that Hegel's *Logic* appears to be a projection of the dialectic of capital into the realm of philosophic categories in which capital becomes deified as Absolute Reason. Whatever conclusions one draws from this close correspondence, it cannot be doubted that by making capital's dialectical logic explicit, Sekine has been able to reconstruct Uno's basic theory of capital in a way that is clearer, more rigorous, and more precise. Furthermore, Sekine's knowledge of mathematical economics enables him to introduce even greater precision into the argument. The combination of dialectics and mathematics is very powerful. For example, the self-conscious use of dialectics enables Sekine to shed new light on what is meant by the transformation of values into prices, and the use of mathematics enables him to show that once we reach the appropriate point in the dialectic of capital, the theory of prices and values are quantitatively determined simultaneously.

Because Sekine has used dialectics to further expose capital's inner identity, we now have a better grasp of exactly what capital is. Hegel's *Logic* (1969) begins with the contradiction between Being and Nothing; whereas the dialectic of capital begins with the contradiction between value and use-value. Hegel's *Logic* proceeds by demonstating how Being overcomes all of the obstacles posed by Nothing until Nothing becomes comfortably internalized in a Being which fully realizes itself as Absolute Reason. Something similar happens with value and use-value, except that in the dialectic of capital we recognize that value is considerably assisted because of an assumption that capital is called upon to produce only ideal and abstract use-values that capital as the motion of value can manage by itself. But since some use-values, such as labour-power, are capable of fighting back, and other use-values such as the steel industry arise contingently in history, the dialectic is disrupted and transformed at more concrete levels of analysis. Thus, at the level of stage theory we cannot expect Capital's dialectical logic always to prevail over use-value and human reistance. Instead it theorizes a stage-specific type of capital accumulation in which value is expected to confront a whole range of more concrete and therefore more recalcitrant use-value obstacles which often can only be successfully managed by capital with the extra-commodity-economic support of ideology, law, and politics. Thus, Sekine's use of dialectics, helps to clarify capital's logic, the resistance to that logic, and the way in which the different levels might be conceived to interrelate.

Since the basic contradiction of the dialectic of capital is between value and use-value, Sekine argues that capitalism's historical stages should be primarily differentiated according to the varying stage specific ways in which the motion of value and its supports manage the most dominant and characteristic type of use-value production in a historical period. Thus, the putting-out production of wool is the dominant type in the stage of mercantilism, the factory-production of cotton in the stage of liberalism, and the corporate-production of steel in the stage of imperialism. Each form of use-value production is commodity-economically managed by a qualitatively different organization of capital: putting-out for wool, competitive light industry for cotton, and oligopolistic corporations for steel. Sekine therefore periodizes capitalist development in accordance with the qualitatively different ways that value expansion manages qualitatively different types of use-value production and their accompanying technological complex.

Up until recently Unoists focused almost entirely on the most abstract level of economic theory. This produced many interesting debates on value theory, but it did not address the difficult and intriguing problems associated with developing the more concrete levels of analysis and their interconnections. Given that most Japanese Unoists are housed in departments of economics and modern academia is typically divided into hide-bound disciplines, there are good sociological reasons why it would be difficult for Japanese Unoists to break out of their preoccupation with abstract economic theory. Also, since Uno had little to say explicitly on the topic of epistemology, some Unoists have drawn from other schools of thought such as regulation theory to fill the perceived vacuum.[6] Finally, the development of stage theory and historical analysis as levels of theory that are distinct, yet informed by capital's logic is intrinsically difficult. For these and other reasons, it has not been easy for Japanese Unoists to develop their own distinctive epistemology and concomitantly their own substantive and distinctive contributions to stage theory and historical analysis.

The Canadian Uno School has at least begun to make some initial contributions in these neglected areas. This has been facilitated by the broad interdisciplinary background of many of those who have learned of Uno's ideas by working with Sekine in Canada, and, of course, by the epistemological character of Sekine's philosophical economics. Indeed, the broad interests of Canadian Unoists is evidenced by the contributions to this book.

This book is divided into two sections that reflect the central preoccupations of Canadian Unoists. Part I focuses on issues and debates at the level of capital's basic operating principles. It is at this level that Uno and most of his Japanese followers have made their greatest contributions. This is also the level at which Sekine has made his signal contribution. Because of his explicit use of dialectics and his knowledge of mathematical economics, Sekine's version of the basic principles of capital is arguably the most rigorous and sophisticated written up to now.[7] In this section of the book there are two contributions by Japanese Unoists, who only came to Sekine's work after learning Uno theory in Japan. Their contributions are interesting because they indicate some of the main lines of debate within Japan and they indicate how Sekine's version of the basic principles fits in with these debates.

Part II is entitled "Philosophical Economics". In comparison with the discipline of mainstream economics which tends to be an unphilosophical "normal science", we see Unoist political economy as highly philosophical. This means to us, in the first instance, that we are self conscious about the unique character of capital as an object of knowledge, about the difficulties of relating abstract levels of theory to more concrete levels, and about the difficulties of relating the economic to the non-economic. These philosophical concerns, which might be said to fundamentally problematize capital as an object of knowledge, are taken up in diverse ways within this book.

It is rather banal to speak of the crisis of Marxism since almost everyone has been speaking about it for the past ten to fifteen years. Because Marxism is seen to be compromised by economism, it is now fashionable to become at least a "post-marxist" if not an "anti-marxist". And yet, Marx's *Capital* is an immense step forward in our understanding of the basic operating principles of capital. Surely the anti-capitalist left should build on this theory. Given the ever-deepening crisis of global capitalism, it seems rather foolish for the left to abandon the terrain of economic theory in the name of moralistic anti-economism or anti-essentialism. And yet how are we to build on the power of Marx's achievement without falling into the sort of economism that has in the past often ignored issues of race, gender, sexual orientation, ethnocentrism, ecology, and neocolonialism?

We think the approach to Marxian political economy outlined in this book offers great promise in answering this question. A theory of a purely capitalist society that clarifies capital's basic operating principles can help us to secure the terrain of economic theory for the left. The idea of levels of analysis opens the possibility of exploring this logic in connection with all other arenas of social life in such a way that there need be no presupposition that capital's logic is always in the last instance *the* determining factor in social life. Indeed, as we move to more concrete levels of analysis, the logic of capital is more and more interfered with and hence our theory becomes less and less of a logic. For example, as noted previously, at the level of stage theory, capital's logic is translated into types of institutional configuration with characteristic dynamics and is no longer expected to achieve dialectical closure that overcomes human resistance. At the level of historical analysis, we cannot determine in advance the role that capital plays in a particular historical outcome without carefully considering all the major social forces that are at work in connection with the outcome under consideration. It is apparent, then, that Unoist political economy offers the possibility of a rigorous theory of capital's "essence", while at the same time avoiding the traditional pitfalls of "essentialism" and "economism". Indeed, we agree with many post-modernist critiques of essentialism. Capital, however, is perhaps the only socio-historical object that does have an essence; thus, it is both possible and desirable to theorize that essence. We think Uno offers a way to both theorize capital's essence and avoid the crude reductionisms that tend to follow in the wake of essentialism. To put it somewhat paradoxically, it offers a "non-essentialist theory of essence" (Albritton, 1993).

On the one hand, Unoist political economy understands how it can be that capital is the only social object of knowledge that has an essence, and at the same time, it recognizes limits to capital that capital itself does not recognize. Thus instead of being mesmerized by capital and its market mentality, it becomes possible to assess realistically its particular effectivity in connection with other social forces such as patriarchy. In this way an Unoist approach to Marxian political economy can effectively dialogue with feminist and postmodernist thought which has often been critical of more conventional and economistic approaches to Marxian political economy. Indeed, the more feminists can advance our understanding of patriarchy, the more clearly we can understand the specific interaction of capitalism and patriarchy. And the same can be said for other areas of concern whether dealing with race, colonialism, or heterosexism.

Notes and References

1. For a more extended critique of Althusser's Marxism from an Unoist perspective see Albritton (1986).
2. We are not aware of precise statistics on this issue. This is therefore an impressionistic estimate made by various Japanese scholars that we have talked to.
3. For a good summary of the basic positions of the Uno School see Mawatari (1985).
4. Critiques of Uno theory by Western Marxists generally do not recognize the internal complexity of the Uno School and often treat Itoh and Sekine as if they were speaking with one voice. It is not uncommon to find critiques of the Uno School as a whole that in fact only apply to one branch or another of the school.
5. In the current intellectual milieu all claims to objectivity are highly suspect, and in our view this is healthy. In this case, however, the claim is well-founded because of the peculiarities of capital as an object of knowledge.
6. Aglietta formulated the suggestive concept "homogeneous space of value" in *A Theory of Capitalist Regulation*. This concept is in many ways close to Uno's idea of pure capitalism, but Regulation Theorists have not taken Aglietta's concept seriously and hence have developed a levels of analysis approach that is problematic. Without clarity on the question of value theory, they cannot sufficiently problematize the relation between levels. The result are strong tendencies towards economism and structural functionalism.
7. Sekine has nearly finished a complete revision of his earlier *The Dialectic of Capital* (1984, 1986).

Part I

The Theory of a Purely Capitalist Society

2 Uno School Seminar on the Theory of Value*

Thomas T. Sekine

In the following ten sets of questions and answers, I intend to give a concise account of the theory of value as it is conceived in the dialectic of capital, i.e., the Marxian economic theory which the late Professor Kozo Uno (1897–1977) developed as a dialectical system. Members of the Uno School, may not be unanimous in what they consider to be the gist of Uno's value theory. What follows should, therefore, be understood to represent my own interpretation of it. Hence the indefinite article in this essay's title.

2.1 ARE "VALUE" AND "EXCHANGE VALUE" THE SAME THING?

No. The synonym for "exchange value" is price or value-form, and not "value". Exchange value is an expression of value, but it is not itself value. The latter refers to that property of a commodity which makes it qualitatively the same as, and only quantitatively different from, another commodity. In other words, value refers to the commodity as representing a part of the abstract-general (i.e., mercantile or commodity-economic) wealth of society.

Although at the beginning of *Capital*, Marx follows the classical practice in stating that "value-in-use" and "exchange value" are the two elements of a commodity, he admits the inappropriateness of that usage only a few pages later and recommends the substitution of the word "value" for "exchange value" there.[1] Yet the importance of this correction has been frequently overlooked.

The confusion between value and price is reinforced by the ambiguous orthodox (neoclassical) usage according to which, for example, the phrase "value and distribution" is just another way of saying "the pricing of outputs and the pricing of inputs". Debreu's *Theory of Value* (1959) merely discusses how general equilibrium "prices" are formed. Nowhere is value defined as distinct from price.

Thus, most would-be critics of the *labour theory of value* (which states that "socially necessary labour" forms the substance of commodity-values) are, in fact, unknowingly criticising a *labour theory of prices* (which presumably claims that the equilibrium relative prices of commodities should be proportional to the quantitites of labour embodied in them); an incorrect theory for which the dialectic of capital is not responsible.[2]

What is vital to the make-up of capitalism is that commodities are produced as value, rather than as use-values. That is to say, they are produced as instruments of trade indifferently to their use-values. Indeed the primary sense of the word "value" is simply "indifference to use-value". Whereas commodities are materially heterogeneous as use-values, they are socially homogeneous as value. It is, moreover, not necessary to know what the substance of this qualitative uniformity is, in the first instance, in order to demonstrate that it gives rise to economically meaningful prices.

All one needs to know is the following. No commodity can express its value except by means of a given quantity of the use-value of another commodity, i.e., by a price. The fact that all capitalistically produced commodities should have an economically meaningful price stems from the fact that they are value-objects.[3]

It is true, of course, that the commodity-form can occasionally (i.e., by chance) attach to such things as works of art, personal services, and even a person's honour and pride, giving them "prices" which express no (economic) value. Such prices are, however, arbitrary, since they do not reflect the underlying allocation of society's resources, which equilibrium prices of genuine commodities do.

Value is thus a property of the genuine commodity produced indifferently to its use-value, and reflects the degree to which its provision costs society in real terms. Only at a later stage of analysis can we explain "in which real terms", however. For the mere observation of commodity exchanges does not enable us to specify the substance of value as labour, or as anything else for that matter.

Marx convinced no one with his immediate deduction from "1 quarter corn $= x$ cwt. iron" that both sides of the exchange equation must contain an equal amount of productive labour.[4] What the equation shows is merely that both sides are equally priced in trade, without any necessary implication that they are of equal magnitude in value. The equation rather implies that the two totally distinct use-values carry a socially uniform quality, the "third" factor, which may be called either value or "moneyness" (convertibility into money), and that this latter gives rise to economically meaningful prices for all commodities.

2.2 WHAT DOES THE VALUE-FORM THEORY DO?

Let me repeat that a commodity can express its value or "money-ness" only in terms of a definite quantity of the use-value of another commodity, which is called an "equivalent". By virtue of the prior expression of value by the first commodity, the equivalent acquires the immediate purchasing-power over it within the terms of the *proposed* exchange. Indeed, the value expression amounts to nothing other than a trade proposal.

For example, in the simple value-form: "20 yards of linen (→) 1 quarter corn", the value of linen is expressed in corn, the equivalent commodity. The sign (→) here should be read "are yours for". This indicates the willingness of the owner of linen to part with its 20 yards, so long as 1 quarter of corn is obtainable. Hence 1 quarter corn is already "little money" which can buy 20 yards of linen from its owner at a moment's notice. The "money-ness" of linen is thus revealed "accidentally", when its owner renders corn "little money" with the qualitative and quantitative restrictions, as per his trade proposal.

What the value-form theory does may be said to remove such qualitative and quantitative restrictions from the "little money" so chosen by the linen-owner, and to show that one particular commodity always in the end emerges as the general equivalent or money, in terms of which all other commodities express their values. With the exclusion of one particular commodity as money (which no longer expresses its own value in terms of another commodity), and with the consequent pricing of all other commodities in its terms, the value expression is completed as: "all the available stock of my commodity (→) whatever it is worth in money", i.e., as the money form of value.[5]

What this statement means is that the necessity of commodities to express their "immanent" value calls money into being. Money here is the universal value-reflecting object, an "external" form of value, i.e., an external object which directly represents value.

We can, therefore, say at this stage what the value of a commodity is like (at least, in the mind of its owner). It is like a given quantity of gold, if gold is the commodity universally adopted as money. Concretely, if I have 1,000 pencils for sale and I price them at 50¢ (in gold) apiece, this means that the value of my merchandise has, in my mind, transformed itself into a $500 worth of gold. Thus the value-form theory makes value, an intangible property of the commodity, tangible, as a given amount of the monetary commodity such as gold. In the present context, however, money is still imaginary (in the mind of the seller) rather than real (in the hands of the purchaser). No physical money need exist merely to price commodities in its terms, that is to say, to put a price-tag on them.

2.3. WHAT DOES THE "MEASURE-OF-VALUE FUNCTION OF MONEY" MEAN? DOES ONE "DETERMINE" VALUE WHILE "MEASURING" IT?

All commodities appear in the market with supply prices, which their sellers (owners) quote tentatively in expressing their values. Since these are money prices, owners of physical money (which is the means of purchase) can buy any commodity offered in the market at will.

The suppliers of the commodity, by pricing it, can only "propose" a trade which they themselves cannot enforce. They can only revise the price, depending on the response of the market, so as to be able to sell their commodity as quickly and as dearly as possible. The owners of money, representing the force of demand in the market, buy immediately if the price is low and temporise if the price is high. This fact compels the sellers increasingly to adjust their supply prices to the demand prices. Therefore, the equality of demand and supply emerges sooner or later, as the commodity is *actually* bought and sold a great number of times in recurrent trade. It emerges, in other words, as money *measures* the value of the commodity by its repetitive purchases, establishing a normal price that measures its value.

The measurement of value thus occurs, when physical money acts as the means of purchase. The price at which the commodity is sold varies from time to time, but eventually settles to a normal level, a level that measures its value. Since a genuine commodity is reproducible, that is to say, it can be supplied in any quantity (in any number of interchangeable samples), its purchase will be repeated until its price settles to a normal level.

What I call a normal price here is, in fact, the same thing as an equilibrium price. The equilibrium price of a commodity in the capitalist market, however, cannot be adequately specified until the real conditions of production, which underlie its demand and supply, are made more explicit. That is why the dialectic speaks of a "normal" price rather than of an "equilibrium" price in this present, abstract context of simple circulation which makes no explicit reference to production. For the same reason, I say that a normal price "measures" the value of a commodity, instead of saying that an equilibrium price "determines" the value of a commodity.[6]

Such dialectical subtlety notwithstanding, the measurement of the value of a particular commodity implies, in the background, the allocation of society's resources for its production relative to that for the production of gold (the monetary commodity). If the commodity is traded at its normal price, that fact implies that it is neither overproduced nor underproduced, and that its production consumes (uses up) just the right quantities of society's resources. It is in this sense that the normal price is said to measure the value of the commodity.

Yet no assertion has so far been made, nor is it generally true, that commodity-values are proportional to normal prices. Suppose that there are two commodities **A** and **B** whose normal prices are $1 and $2 respectively. This does not imply that, if **A**'s value is x, **B**'s value ought to be $2x$ (for some positive number x). In fact, **B**'s value may well be only $0.5x$. No contradiction exists between the two statements: (1) that only when **B**'s price is twice as much as **A**'s are these commodities produced in the right quantities; and (2) that one unit of **B** then consumes half as much of society's resources as one unit of **A** does. This assertion leads us to the definition of the substance of value.

2.4 WHAT IS THE SUBSTANCE OF VALUE? CAN THERE BE A NON-LABOUR THEORY OF VALUE?

Capitalism differs from "simple commodity production" in, among other things, that the former, by virtue of the conversion of labour-power into a commodity, is necessarily indifferent to the production of specific use-values. Unlike the small commodity producer, the capitalist has no stake in the production of particular goods. Capital, in other words, produces *any* commodity if it is profitable to do so (just as money buys *any* commodity if it is desirable to do so). It amounts to saying that capital produces commodities as *value*, i.e., with indifference to their use-values.

The production of commodities as value which characterises the capitalist economy, however, depends on the conversion of labour-power into a commodity. For only commoditised labour-power can be indifferent to the use-values that it produces.

Labour that produces a use-value (a material object for use or consumption) is called productive. Labour-power is the capacity to perform productive labour which, as is well known, is both concrete-useful and abstract-human. The classical school understood this duality of productive labour only vaguely, which made their labour theory of value inconclusive. It was with the explicit recognition of this duality that Marx removed all ambiguities from the labour theory of value. The fact that in all societies productive labour has the abstract-human aspect makes labour-power, in principle, a non-specific factor of production.

This means that, at some finite cost, human labour can always be shifted from one concrete-useful form to another. There may be cases in which the cost is prohibitive, but the capitalist method of production tends to reduce that cost, if not to eliminate it altogether, by the simplification of labour. It is precisely for this reason that capital which purchases labour-power as a commodity can produce *any* commodity with indifference to its use-value, while labour, on the other hand, becomes an abstract "disutility" to the worker regardless of which concrete-useful form it is rendered in.

Now it is obvious even to a child that there are no such things as "abstract-spatial land" and "abstract-physical capital". Land, though an original factor of production, comes only in various concrete-useful forms (in location, fertility, etc.); moreover, the production of use-values does not "consume (use up)" land which is a free gift of nature to society.[7]

Capital, in the sense of capital-goods, is not even an original factor of production. Moreover, the fact that every item of capital is a specific assemblage of various elements of production (not unlike the pieces of a meccano set?) explains the inveterate non-maleability and immeasurability of capital. This characteristic of capital caused a great uproar in bourgeois economics (and even led to the invention of that prismatic "standard commodity", which not only fascinates but also spellbinds some comprador Marxists).

Clearly, both a land theory and a capital theory of value are idle bourgeois fantasies with no credible economic foundation.[8] Productive labour alone, which flows from labour-power, forms the substance of value. For labour-power is the only element of production which is both original and non-specific.

When commodities are produced in socially necessary (equilibrium) quantities, i.e., are neither overproduced nor under-produced, we say that "socially necessary labour" has been spent for their production.

Hence a precise statement of the labour theory of value is that socially necessary labour forms the substance of commodity values. Indeed, the capitalist production of commodities as value means nothing other than the production of commodities as embodiments of socially necessary labour. This fact also implies that the magnitude of the value of a commodity is defined (determined) by the quantity of socially necessary labour embodied in it. Socially necessary labour is the only real cost that society incurs in the production of commodities.

2.5 WHAT IS THE NECESSITY OF THE LAW OF VALUE?

A close relation has already been observed between the existence of capitalism (the conversion of labour-power into a commodity) and the validity of the labour theory of value (the production of all commodities as embodiments of socially necessary labour). The law of value says that, when the labour theory of value holds, capitalist society is viable, and vice versa. A society is said to be viable if its direct producers have a guaranteed access to the product of their necessary labour, since with it they can reproduce their labour-power.

Thus capitalism is viable as an historical society, if and only if the money wages paid to workers enable them to "buy back" enough wage-goods for the reproduction of their labour-power. We have to make sure that this condition of viability is consistent with the labour theory of value, i.e., with the capitalist indifference to the production of use-values. The necessity of the law of value means that the viability of capitalism implies and is implied by the validity of the labour theory of value.

Let us assume the following: (1) all workers work for 12 hours a day with the rate of surplus value of $e = 1$; (2) 3 units of **A**, where **A** represents the basket of assorted wage-goods, suffice to reproduce the daily consumption of labour-power; and (3) no fixed capital exists. With these general assumptions consider the following specific case. A capitalist produces, per employment of one worker, 15 units of **A** in a day with the rate of profit of $r = 0.2$; the value and the price of **A** per unit are respectively $\lambda_a = 2$ and $\rho_a = 1$.

(A)	c		v		s		
in labour	18	+	6	+	6	= 30	$\lambda_a = 2$
in money	9.5	+	3	+	2.5	= 15	$\rho_a = 1$
in quantity	9.5	+	3	+	2.5	= 15	$r = 0.2$

Then we are in effect confronted with the situation described in the above table. Obviously the capitalist must pay $3 as the money wage so that the worker can, with it, buy back 3 units of **A**, which are the product of his necessary labour of 6 hours.

In order for the labour theory of value to hold, so that only socially necessary labour is applied for the production of all commodities, the latter must, of course, be produced in socially necessary quantities. This further implies that the production of all commodities must be equally profitable, and that capitalists are, therefore, indifferent to the use-value aspect of production. So let **B** whose value happens to be $\lambda_b = 1$ be any commodity different from **A** and producible with the rate of profit of $r = 0.2$. Suppose now that another capitalist advances, per employment of one worker, $9 for the daily production of 24 units of **B**. Then from the information available, we can construct the following table:

(B)	c		v		s		
in labour	12	+	6	+	6	= 24	$\lambda_b = 1$
in money	6	+	3	+	1.8	= 10.8	$\rho_b = 0.45$
in quantity	13.33	+	6.67	+	4	= 24	$r = 0.2$

Here again the worker must be paid $3 with which to buy 3 units of **A** for the reproduction of his labour-power, and that is done. Since **B** represents *any* commodity other than A, it follows that capitalism is viable, when all commodities are produced with only socially necessary labour. This establishes the necessity of the law of value. For, even though prices are not proportional to values, there clearly exist some prices that ensure both the validity of the labour theory of value and the viability of capitalist society.

That, however, is quite different from claiming a theory of the (quantitative) determination of values and prices. Here we have assumed (λ_a, λ_b, e) and (ρ_a, r), together with various parameters, and deduced only ρ_b, although this latter represents all "other" prices. The implication is that a set of equilibrium prices relative to ρ_a and r exists which appropriately allocates socially necessary labour to the production of all commodities, so as to make capitalist society viable.[9] For the theory of (quantitative) determination of values and prices, the nature of productive techniques must be known more explicitly.

2.6 HOW DO YOU QUANTITATIVELY DETERMINE VALUES AND PRICES?

In the appendix to this paper I provide the algebra for the simplest case consisting of one capital good, one wage-good, and one luxury good (consumption good for capitalists). Here I would rather illustrate the theory with concrete numbers. Suppose that 150 units of the capital good are produced with 50 units of the capital good itself and 20 hours of labour. Let us write this technical and allocational relation as $(50, 20) \rightarrow 150$. We shall adopt the same convention for the other two goods, and suppose the following technology complex (T) to correspond to system (1) in the appendix:

		capital-good	labour		output		
		(50,	20)	→	150	=	X
(T)		(40,	30)	→	80	=	Y
	+)	(30,	40)	→	90	=	Z
		120,	90.				

For the system to be meaningful, we must first ascertain that the amount of the capital-good produced (150 units) is not smaller than the amount of it used up (120 units). This condition of self-replacement or productiveness, which in the appendix appears as a weak inequality (2) is the same thing as what Uno calls "the absolute foundation of the law of value". The operation (T), shown as (3) in the appendix, then enables us to find values which in the present case will be $(\lambda_x, \lambda_y, \lambda_\zeta) = (0.2,\ 0.475,\ 0.511)$.

To determine the rate of surplus value we must make use of the identity: $\lambda_y Y = 90/(1+e)$, which corresponds to (4) in the appendix. This identity says that the workers "buy back" the output of the wage-good with their necessary labour, a proposition already referred to in connection with the "necessity of the law of value". In the present case we find a positive rate of surplus value, $e = 1.368$. Since a non-positive rate of surplus value is meaningless, T must satisfy the inequality $\lambda_y 80 < 90$ to be capitalistically operable. This and the condition of self-replacement, however, in effect guarantee positive prices and a positive rate of profit.

Now consider the operation P(T), as shown by (5) in the appendix. This system of three equations is clearly under-determined. For even when we assume an arbitrary wage-rate (such as $w = 1$), we still have three prices (ρ_x, ρ_y, ρ_z) and the rate of profit (r) to determine. I, therefore, introduce the identity $\rho_y Y = w90$, corresponding to (6) in the appendix, which says that the money value of the output of the wage-good is entirely bought by the total wage-bill paid during the period. This identity, which I call "the fundamental constraint of the capitalist market", is clearly a market expression of the viability condition of capitalism. (Hence, unlike Walras' Law, this constraint is not arbitrarily imposed as a postulate. It is derived from the previous theory of the necessity of the law of value.) Prices and the rate of profit can now be calculated, when $w = 1$, as $(\rho_x, \rho_y, \rho_z) = (0.5477, 1.125, 1.087)$ and $r = 0.7338$.

The present theory can be easily generalised to apply to a system containing any number of commodities. It can also allow for fixed capital and differences in the turnover-frequencies of capital. At this point, however, I simply wish to call attention to several points of theoretical importance, a full discussion of which would require more space. (i) All commodities can be categorised into the three groups. (ii) The technology complex is specific to the indicated activity levels, implying not only technology in the narrow sense but also an equilibrium allocation of capital. (iii) The outputs (150, 80, 90) emerge at the end of the period, but the inputs (120, 90) must be there at the beginning. Take note, however, that variable capital ($w90$) is "advanced" only in money terms, not in physical terms as a stock of wage-goods. (iv) A technology complex which does not satisfy the condition of self-replacement with respect to all capital-goods and the condition that $e > 0$ cannot be capitalistically operated. (v) The method of the fundamental constraint is crucial to the solution of the price system, especially when there are more than one wage-good.[10]

2.7 WHAT IS THE TRANSFORMATION PROBLEM? WHAT IS ITS ECONOMIC SIGNIFICANCE?

I would like to emphasize that there are two transformation problems. One involves the dialectical (i.e., qualitative or conceptual) transformation of values into prices and of surplus value into profit. Here there is no question of an inverse transformation. The other has to do with a mathematical (i.e., quantitative or formal-logical) transformation of values into prices, and of the *rate of* surplus value into the *rate of* profit. In this case the inverse transformation is part of the problem.

In the dialectic, which is the logic of synthesis, the same concept reappears a number of times but each time it is increasingly more specified or concretised, and therefore assumes different names. What used to be called *A* at a more abstract (i.e., less synthetic) level of discussion now appears as *B*, redefined to fit the new context. When this sort of thing occurs we speak of the conversion (or transformation) of *A* into *B*. For example, "the conversion of the value of labour-power into wages" means that what was referred to, in earlier contexts, as the value of labour-power is now specified as money-wages. "The transformation of surplus profit into rent" means that surplus profit which arises in a particular context, i.e., specifically in relation to the differential fertility of land is called rent. There are, of course, many other instances of such usage in *Capital.*

Such transformations are like that of Cinderella into the Crown Princess. The person has not changed but the context has, requiring her to adjust her appearance and demeanour accordingly. In the dialectic of capital the transformation of values into prices occurs, when one moves from the doctrine of production, in which the production of commodities is strictly viewed as the production of value, to the doctrine of distribution, which takes the distinctness of use-values in the capitalist market into explicit consideration. This statement means that values and prices, which could not be quantitatively determined in the earlier context, may now be completely spelled out, as capital emerges from the underworld of production to the limelight of the capitalist market, permitting the full specification of the technology complex (T).

In the following chart the lateral arrow indicates the qualitative transformation. Prices that remained in the shadow of values in the doctrine of production come to the forefront in the doctrine of distribution, relegating values to the background. That is why we speak of $\bar{\Lambda} \rightarrow P(T)$ rather than $\bar{P} \rightarrow \Lambda(T)$. But the real issue here is the transformation of quantitatively unspecified values and prices into them so specified.

	Doctrine of Production		Doctrine of Distribution	
Explicit	$\Lambda = \bar{\Lambda}$	$\rightarrow$	$P = P(T)$	$\uparrow\downarrow$
Implicit	$P = \bar{P}$		$\Lambda = \Lambda(T)$	

The word "transformation", however, is not monopolised by the dialectic. When it is used in mathematics (as, e.g., linear, power, Laplace transformation, etc.), it means a transplantation of a mathematical point or relation from one space (or system of coordinates) to another. Similarly, we can speak of the mathematical transformations of values into prices and of prices back into values. These are shown by the pair of vertical arrows in the above chart. That is to say, if we first know values and the rate of surplus value, we can get to the corresponding prices and rate of profit, provided that we also have certain information about the value compositions of capital. And under similar conditions we can take the reverse course as well.

To see this, rewrite the first two equations of $\Lambda(T)$ and $P(T)$, assuming $w = 1$, respectively as follows:

$$\begin{cases} \lambda_x 150/20 = k_x/(1+e) + 1, \\ \lambda_y 80/30 = k_y/(1+e) + 1, \end{cases} \qquad (*)$$

and

$$\begin{cases} \rho_x 150 = 20(1+r)(k'_x + 1), \\ \rho_y 80 = 30(1+r)(k'_y + 1), \end{cases} \qquad (**)$$

where $k_x = \lambda_x 50(1+e)/20$ and $k_y = \lambda_x 40(1+e)/30$ are the value compositions of capital, whereas $k'_x = \rho_x 50/20$ and $k'_y = \rho_x 40/30$ are the corresponding price compositions of capital. Clearly,

$$k'_x/k_x = k'_y/k_y = \rho_x/\lambda_x(1+e).$$

Moreover, $\lambda_y(1+e) = \rho_y$ from (4) and (6) of the appendix. Hence, if we know (e, λ_x, k_x, k_y), we can always derive (r, ρ_x); and if we know (r, ρ_x, k'_x, k'_y), then we can get (e, λ_x). Here I illustrate just this latter transformation. Write $(*)$ above as:

$$\lambda_x 150 = 20(k'_x\lambda_x/\rho_x + 1), \ \lambda_y 80 = 30(k'_y\lambda_x/\rho_x + 1),$$

and recall that $(r, \rho_x, k'_x, k'_y) = (0.7338, 0.5477, 1.369, 0.7303)$. Then we get $\lambda_x = 0.2$ and $\lambda_y = 0.475$. Since the second equation is equal to $90/(1+e)$ by the viability condition of capitalism, we also get $e = 1.368$. The same results have been obtained from the direct solution of $\Lambda(T)$.

The meaning of this transformation is self-explanatory. It establishes that one point $(\lambda_x, \lambda_y, \lambda_z, e)$ in the value space has a unique corresponding point $(\rho_x, \rho_y, \rho_z, r)$ in the price space and *vice versa*, just as one point in the Cartesian coordinate system has its image in the polar coordinate system and *vice versa*. The same mathematical relations hold differently in different spaces.

2.8 WHAT IS THE LAW OF AVERAGE PROFIT?

When values and prices are quantitatively determined, it turns out that they are not, in general, proportional to each other. Yet the deviation of equilibrium prices from values is by no means arbitrary. The law of average profit claims that the extent to which equilibrium prices (production-prices) depart from value-proportional prices is strictly predetermined by the variability of techniques in the production of individual commodities as use-values, so that equilibrium prices are, as it were, tethered to values. In other words, the law of average profit defines the concrete mode of enforcement of the law of value through the motion of prices in the capitalist market.

The law of average profit, however, is often phrased in a somewhat more specific form, such as: "If it is assumed that the price of the aggregate social product is equal to its value, then the production-price of the commodity produced with a higher-than-the-social-average value composition of capital exceeds its value-proportional price, and *vice versa*." This statement applies strictly to the case in which (1) fixed capital does not exist, and (2) there is no more than one capital-good in the system. In other cases small modifications are necessary. I shall refer to this proposition as the *first law of average profit*.

In chapter XI of *Capital*, Volume III, Marx also advances a closely related proposition that: "When wages are raised, the production-price of a commodity produced with a higher(lower)-than-the-social-average value composition of capital falls (rises); and when wages are lowered, the exact reverse occurs." I refer to this latter proposition as the *second law of average profit*.

It can easily be established that, if the rate of surplus value is assumed to be zero (i.e., $e = 0$), i.e., if no surplus labour is performed, values and prices are proportional. Such a case is, of course, capitalistically meaningless, since a zero rate of surplus value would mean a zero rate of profit (by virtue of the so-called fundamental theorem). However, real wages are at their "theoretical" maximum when the rate of profit is zero and prices are consequently proportional to values. Now from this limiting point, let wages fall, and profits emerge. Prices will then begin to diverge from values; and do so the more, the more the rate of surplus value increases. This is the meaning of the second law of average profit. In fact it combines the first law of average profit with the well-known Ricardian theorem that, given the technology, a gain in the rate of profit is a loss in real wages, and *vice versa*. (The same idea is embodied in the so-called "factor-price frontier" in neo-classical economics.)

In the previous example, we had $(k_x, k_y, k_z) = (1.1842, 0.6316, 0.3553)$. Let us define the social-average composition of capital as $k = (20k_x + 30k_y + 40k_z)/90$, using labour-inputs for weights. It turns out that in this particular case $k = 0.6316$ is also, by chance, equal to the value composition of capital in the wage-good producing sector ($k_y = 0.6316$). On the other hand, in order to equate total value ($\lambda_x X + \lambda_y Y + \lambda_z Z = 114$) to total price ($\rho_x X + \rho_y Y + \rho_z Z = 270$), let us apply the conversion rate $\alpha = 2.3683$ to labour values (λ_i) and obtain value-proportional prices ($q_i = \alpha\lambda_i$), which are $(q_x, q_y, q_z) = (0.4737, 1.1295, 1.2100)$.[11]

It is now possible to compare equilibrium prices and value-proportional prices, and derive the following conclusion:

$$\rho_x = 0.548) > q_x(= 0.474), \text{ since } k_x(= 1.1843) > k(= 0.6316)$$

$$\rho_y(= 1.125) = q_y(= 1.125), \text{ since } k_y(= 0.6316) = k(= 0.6316)$$

$$\rho_z(= 1.087) < q_z(= 1.210), \text{ since } k_z(= 0.3551) < k(= 0.6316)$$

This illustrates the first law of average profit.

In order to illustrate the second law, I let the production of the wage-good increase from $Y = 80$ to $Y' = 85$. For simplicity, however, I assume: (i) that this is entirely at the expense of the luxury-good and does not affect the output level of the capital-good; (ii) that the total number of hours of current labour is unchanged at 90; and (iii) that the change in the pattern of social demand does not affect the methods of production. We then have the new technology complex:

$$(\text{T}') \left\{ \begin{array}{rllcll} & (\ 50, & 20\ ) & \rightarrow & 150 & = X' \\ & (\ 42.500, & 31.875\ ) & \rightarrow & 85 & = Y' \\ +) & (\ 28.594, & 38.125\ ) & \rightarrow & 85.781 & = Z' \\ \hline & 121.094, & 90 & & & \end{array} \right.$$

which satisfies the condition of self-replacement. In this case we have $(\lambda_x, \lambda_y, \lambda_z, e) = (0.2, 0.475, 0.511, 1.229)$, so that values are unchanged but the rate of surplus value has fallen (real wages have risen). The value compositions are, of course, uniformly lower by about 5.87% to reflect the change in the rate of surplus value: $(k_x, k_y, k_z) = (1.1148, 0.5945, 0.3343)$, and their labour-weighted social-average changes to $k = 0.5999$, which is only about 5.02% lower than before. By solving the price equations, we find out that $(\rho_x, \rho_y, \rho_z, r) = (0.5098, 1.0588, 1.0327, 0.6810)$, assuming again that $w = 1$. In this case, we should have $\alpha = (\rho_x X' + \rho_y Y + \rho_z Z)/(q_x X + q_y Y + q_z Z) = 255.06/114.22 = 2.2373$, which implies $v = \alpha/(1 + e) = 1.0038$.[12]

Now, according to the second law of average profit, the fall in the rate of surplus value from 136.8% to 122.9% (which raises v from 1 slightly to 1.0038) should make prices more proportional to values. That is to say, we expect that ρ_x/ρ_y falls towards q_x/q_y since $kx > k$, and that ρ_z/ρ_y rises towards q_z/q_y since $k_z < k$. Indeed ρ_x/ρ_y which used to be 0.4868 has fallen to 0.4815, coming closer to $q_x/q_y = 0.4210$; and ρ_z/ρ_y which used to be 0.9662 has risen to 0.9753, coming closer to $q_z/q_y = 1.0779$. The value ratios q_x/q_y and q_z/q_y are, of course, unchanged by the fall in the rate of surplus value, since the methods of production (technology in the narrower sense) are assumed to be the same.

2.9 HOW DOES THE LAW OF MARKET VALUE SUPPLEMENT THE LAW OF AVERAGE PROFIT?

So far it has been supposed that there exists only one technique for the production of a particular commodity. This assumption is surely unrealistic. A use-value is often produced with different techniques, so that in the same industry some firms earn more, and others less, than the average profit. In other words, positive or negative surplus profits are the norm. This multiplicity of techniques, however, does not invalidate the labour theory of value, but we must now take into account the principle of market value, which enables us to identify the value (or equilibrium price) determining technique, or combination of techniques, from among those actually employed by different firms in the industry.

Suppose that, for the production of steel ($X = 150$) with itself and labour, two techniques are simultaneously used as follows, their outputs being $X^{(1)} = 80$ and $X^{(2)} = 70$, respectively.

$$
\begin{array}{lllll}
 & (30, & 8) & \rightarrow & 80 = X^{(1)} \\
 & (10, & 20) & \rightarrow & 70 = X^{(2)} \\
\hline
+) & (40, & 28) & \rightarrow & 150 = X
\end{array}
\qquad (***)
$$

Hence *in toto* 150 units of steel are produced with 40 units of steel itself and 28 units of labour. But $(40, 28) \rightarrow 150$ cannot immediately be used to determine the value of steel. In the presence of two (or more) techniques, we cannot know in advance which technique or combination of techniques will be used to respond to an autonomous change in the demand for steel.

When the demand for X increases, it may be $X^{(1)}$ or $X^{(2)}$ or some combination of both that supplies the required addition. Let us suppose that when the demand for X increases by 10 units, $X^{(1)}$ increases by 8.205 units and $X^{(2)}$ by 1.795 units. This means that when X must increase overall by 6.667%, $X^{(1)}$ increases by 10.256% and $X^{(1)}$ by 2.564%, so that the elasticity of $X^{(1)}$ at the margin is $\delta^{(1)} = 1.5384$ and the corresponding elasticity of $X^{(2)}$ is $\delta^{(2)} = 0.3846$. The two techniques shown in (***) above must be weighted by these elasticities before being added up to the synthetic value-determining social technique as:

$$
\begin{array}{lllll}
 & (46.16, & 12.31) & \rightarrow & 123.08 \\
 & (\ 3.84, & 7.69) & \rightarrow & 26.92 \\
\hline
+) & (50, & 20) & \rightarrow & 150.
\end{array}
$$

All techniques in (T), as shown earlier in my answer to question 6, are social techniques synthesised in the same fashion.

Refer to my paper on market value published elsewhere for a more detailed account of the same principle, including its possible application to the cases of heterogeneous labour and joint production.[13] Here I wish to reaffirm that values and equilibrium prices are meaningful concepts, if and only if resources readily flow *at the margin* of all capitalist industries. Thus if, as demand shifts from industry **A** to industry **B**, enough resources migrate from **A** to **B** so as to adjust the output levels to the altered conditions of demand, then the law of value is securely preserved. It is by no means necessary that all currently employed resources should actually move between **A** and **B**.

The development of capitalism, with its attendant simplification of labour, ensures an adequate (but never complete) mobility of resources for the marginal adjustment of industrial outputs. The law of value works, even if immobile and non-competing elements persist in the interior of various industries. Such elements give rise to "false social value" and surplus profits (positive or negative) that express it. The law of value that takes such contingent matters into consideration is called the law of market value. It, together with the law of average profit, describes the concrete mode of enforcement of the law of value.

2.10 HOW DOES THE PRESENCE OF LANDED PROPERTY BEAR ON THE REGULATION OF CAPITALIST SOCIETY BY THE LAW OF VALUE?

In collecting differential rent, landed property merely assists capital rather than interfering with its logic. Let us now reinterpret that the first two lines of (***) above represent the production of wheat with itself and labour per 100 acres of two different types of land. Land-(1) is more fertile than land-(2), since output per worker is only 3 on land-(2) whereas it is 6.25 on land-(1). The differential fertility of land is, however, given by nature, and does not tend to be eliminated by the competition of capital. Hence the market-regulating value and production-price of wheat should be equal to the "individual" value and production-price of wheat harvested on the least fertile land-(2).

If we calculate $(\rho_x/w,\ r)$ with $(30,\ 8) \rightarrow 80 = X^{(1)}$ and $(40,\ 30) \rightarrow 80 = Y$, assuming that the total labour employed in the system is 90, we get $(\rho_x/w, r) = (0.5235, 0.7667)$. If we use $(10, 20)\ 70 = X^{(2)}$ instead, other things being equal we get $(\rho_x/w, r) = (0.6156, 0.6476)$. These latter numbers are respectively the market regulating production price of wheat and the general rate of profit. Hence the farmers who produce on land-(1) earn the surplus profit of $0.6156 - 0.5235 = 0.0921$ per unit of wheat, which should, of course, be converted into differential rent, if the capitalist market is to guarantee fair competition.

If $X^{(1)}$ and $X^{(2)}$ refer to the outputs of the first and the second instalment of investment on 100 acres of the same land, rather than the outputs per 100 acres of different lands, the formal aspects of the problem remain unchanged. The first instalment is more productive than the second, and hence pays a rent proportional to the differential productivity of investment. In collecting differential rent of either kind, landed property has so far not actively interfered with the determination of the rate of profit and production-prices by capital.

The situation changes altogether, when landed property collects absolute rent. By limiting the supply of agricultural land, landed property can raise the price of agricultural products above their production-prices, and thus appropriate part of surplus value produced in agriculture, *before* its capitalist distribution as average profit. Since the value composition of capital in agriculture tends to be lower than the social average, the production-prices of agricultural products are lower than their value-proportional prices, meaning that surplus value normally tends to be transferred from agriculture to non-agriculture.

Now the appropriation of agricultural surplus value by landed property in the form of absolute rent certainly reduces average profit, which is the form of distribution of surplus value to capital. Yet, it does not reverse the flow of surplus value from agriculture (or industries with lower capital composition) to non-agriculture (or industries with higher capital composition).

It is, however, technically possible for landed property to limit the supply of land so drastically as to raise the market prices of agricultural goods above their value-proportional prices. In such a case, landed property would earn monopoly rent in addition to absolute rent, and this would disrupt the operation of the law of average profit. For surplus value would then have to flow from industries with higher capital composition (agriculture) to industries with lower capital composition (non-agriculture), a movement which would be exactly opposite of that established earlier by the principle of the distribution of surplus value as profit, once that principle itself was established by the operation of the law of average profit.

Should this happen, the regulation of capitalist society by the law of value would be suspended, and a Malthusian underconsumption (if not a Ricardian stationary state) could no longer be ruled out as a chimera. Genuine capitalism, however, consists of a "teleological coexistence" of capital and landed property in which the latter, being "satisficed" (an expression due to H. A. Simon to describe the sense of having achieved sufficiency in satisfaction) with increasing wealth, refuses to yield to the bourgeois addiction to maximise revenues.

Appendix

Consider the simplest capitalist economy in which one means of production, or capital good, (*X*), one wage-good (Y), and one consumption-good for capitalists, or luxury good, (Z), are produced and competitively traded. If X_i (i = x, y, z) represent the quantities of the capital-good, and L_i (i = x, y, z) the number of hours of labour spent for the production of X, Y, Z, then the technology complex can be written as follows:

$$(T) \quad \begin{cases} (X_x,\ L_x) \to X, \\ (X_y,\ L_y) \to Y, \\ (X_z,\ L_z) \to Z. \end{cases} \tag{1}$$

The component techniques of T are all value-determining "social" techniques systhesized by the method discussed in Section 9 of the text.

In order for this economy to be reproducible, it is necessary that the condition of self-replacement

$$X_x + X_y + X_z \leq X \tag{2}$$

should hold. This guarantees that all values λ_i (i = x, y, z) are meaningfully positive. The Λ-operation is as follows:

$$\Lambda(T) \quad \begin{cases} \lambda_x X_x + L_x = \lambda_x X, \\ \lambda_x X_y + L_y = \lambda_y Y, \\ \lambda_x X_z + L_z = \lambda_z Z. \end{cases} \tag{3}$$

In order to determine the rate of surplus value (e), the following constraint (the viability condition) must be used:

$$\lambda_y Y \equiv (L_x + L_y + L_z)/(1 + e). \tag{4}$$

This constraint says that the workers are guaranteed to receive the product of their necessary labour.

Write equilibrium prices, or production-prices, as ρ_i (i = x, y, z), the rate of profit and the money wage-rate as r and w. Then the P-operation on T is:

$$P(T) \quad \begin{cases} (\rho_x X_x + wL_x)(1 + r) = \rho_x X, \\ (\rho_x X_y + wL_y)(1 + r) = \rho_y Y, \\ (\rho_x X_z + wL_z)(1 + r) = \rho_z z. \end{cases} \tag{5}$$

These equations together with the constraint

$$\rho_y Y \equiv w(L_x + L_y + L_z) \tag{6}$$

can be solved for an arbitrary w. This "fundamental constraint of the capitalist market" says that the workers do not save, and hence that the total wages bill paid in society should equal the total money value of wage-goods produced in that society. Clearly (6) is the market expression of (4). If, in addition to (2), we also have the condition e 0 in (4), then the prices and the rate of profit are bound to be positive.

Notes and References

* A shorter version of this paper was published in *Science and Society*, XLVIII (1984), pp. 419–32. The present version considerably expands on it, and adds a number of numerical examples.

1. "When, at the beginning of this chapter, we said, in common parlance, that a commodity is both a use-value and an exchange value, we were, accurately speaking, wrong. A commodity is a use-value or object of utility, and a value." *Capital* (the International Publishers edition) I, p. 60.
2. The problem goes back to the classical school, whose labour theory of value was described to be only about 93% a labour theory of *price*. (See the well known article by G. J. Stigler: "Ricardo and the 93 Per Cent Labour Theory of Value", *The American Economic Review*, XLVIII (1958), reprinted in G. J. Stigler, *Essays in the History of Economics*, University of Chicago Press, Chicago, 1965. pp. 326–42. Stigler, of course, is talking about the labour theory of price, not of value, in that essay.) Even Marxists, many of whom fail to distinguish between value and exchange value despite Marx's express warning, have unknowingly upheld a labour theory of *price* in the name of the labour theory of *value*.
3. For the argument that "scarcity" does not explain the cause of prices, see Thomas T. Sekine, "The Necessity of the Law of Value", *Science and Society*, XLIV (1980), p. 294.
4. See Kozo Uno, *Principles of Political Economy*, Humanities Press, New Jersey, 1980, pp. 32–4, note 2.
5. For details see my "Pricing of Commodities", *York Studies in Political Economy*, IV (1985), pp. 97–121.
6. I also alert the reader to another (qualitative) usage of the word "determine (bestimmen)" as in Marx's Expression "Socially necessary labour *determines* the value of a commodity". I take this to mean that "socially necessary labour defines or forms the substance of value", rather than that "the magnitude of value is represented by the number of hours of socially necessary labour".
7. Marx simply accepted the Ricardian view that land was an "indestructible" element of production. More recently, however, we have been made keenly aware that the "topsoil" can easily be destroyed by misuse or lack of preservatory care. In theory, the teleological coexistence of private landed property with capital, makes it impossible for the latter to unilaterally spoliate the productivity of land, so that it is *as if* land were

indestructible in a genuinely capitalist regime. The fact that agribusiness does not recognise the separation of capital from landed property means that it is *not* a genuinely capitalist industry and hence has no relevance to the law of value.

8. P. A. Samuelson, "Wages and Interest: A Modern Dissection of Marxist Economic Models", *The American Economic Review*, XLVII (1957), pp. 181–219. E. D. Domar, *Essays in the Theory of Economic Growth*, Oxford University Press, 1957, p. 26, p. 88.
9. See my "Necessity of the Law of Value, Its Demonstration and Significance" (Chapter 3 in this volume) for more detail.
10. In the absence of this constraint, the solution of the price system requires the unreasonable assumption that the workers are paid in physically prescribed wage-baskets with no option to retrade their contents, i.e., that they are paid "fodder" instead of money wages. Such an assumption which in effect reduces labour-power to an intermediate good, and renders the distinction between "variable" and "constant" capital irrelevant is incompatible with the presuppositions of the dialectic of capital. See Thomas T. Sekine, "The Transformation Problem, Qualitative and Quantitative", *York Studies in Political Economy*, VI (1985), pp. 60–96.
11. Confirm that $\alpha = 2.3683$ makes total value equal to total price, i.e., $q_x X + q_y Y + q_z Z = 270 = \rho_x X + \rho_y Y + \rho_z Z$. The following discussion, however, does not depend on this particular choice of the "postulate of invariance", so called by F. Seton in his "The Transformation Problem", *The Review of Economic Studies*, XX (1957), pp. 149–60. In the present case, it so happens that another invariance postulate $q_y Y = 90 = \rho_y Y$ also holds with the same $\alpha = 2.3683$. That, however, implies that $v = \alpha/(1+e) = 1$, which is the money wage-rate that would prevail if prices were proportional to values.
12. See the preceding note 11.
13. Thomas T. Sekine, "The Law of Market Value", *Science and Society*, XLVI (1983), pp. 420–44.

3 The Necessity of the Law of Value, Its Demonstration and Significance

Thomas T. Sekine

3.1

In paragraph 31 of his *Principles* Uno illustrates the labour-and-production process by means of the following numerical example: 6 kg. of raw cotton, one spinning machine, and 6 hours of labour produce 6 kg. of cotton yarn.[1] It is assumed that no surplus labour is performed. It is also assumed that the 6 kg. of raw cotton embody 20 hours of labour, and the current depreciation of the spinning machine is worth 4 hours of labour. Hence the means of production already embody 24 hours of labour. This and the 6 hours of current labour thus produce the 6 kg. of cotton yarn which embody 30 hours of labour.

Further along in paragraph 34 where he explains the process of value formation and augmentation, Uno supposes that the means of livelihood needed for the reproduction of labour-power per day can be produced in 6 hours of labour, and that the money price of the basket is 3 shillings. When the production of cotton yarn illustrated earlier is to be capitalistically operated, Uno supposes furthermore that the constant capital which embodies 24 hours of labour can be purchased for 12 shillings and the 6 kg. of cotton yarn which embody 30 hours of labour can be sold for 15 shillings. Hence the capitalist recovers the 12 shillings that he spent on the means of production by selling four-fifths (or 4.8 kg.) of his output of yarn, and 3 shillings that he paid the worker by selling the remaining one-fifth (or 1.2 kg.) of the same.[2]

The above information may be tabulated as follows, where (A) refers to the production of the means of livelihood (wage-good) and (B) to that of cotton yarn, per employment of one worker. All numbers in square brackets are supplied additionally by the present writer. Let *u* stand for the product value, and c, v, s, respectively, for its constant-capital, variable-capital and surplus-value component.

(A)	c		v		s		u	
in labour	[18]	+	6	+	0	=	[24]	$\lambda_a = 3$
in money	[9]	+	3	+	0	=	[12]	$\rho_a = 1.5$
in quantity	[6]	+	[2]	+	0	=	[8]	$r = 0$

(B)	c		v		s		u	
in labour	24	+	6	+	0	=	30	$\lambda_b = 5$
in money	12	+	3	+	0	=	15	$\rho_b = 2.5$
in quantity	4.8	+	1.2	+	0	=	6	$r = 0$

From the numbers newly supplied in square brackets it should be inferred that 2 baskets of wage-goods are assumed necessary and sufficient to reproduce a one-day consumption of labour power. The symbols λ_a (ρ_a) and λ_b (ρ_b) denote the values (prices) of the basket of wage-goods and of cotton yarn, respectively; the symbol r denotes the rate of profit.

3.2 THE PROBLEM

Now in paragraph 35, still discussing the process of value formation and augmentation, Uno lets the spinning capitalist double the scale of his operation not only by extending the working-day from 6 to 12 hours but also by applying twice the means of production. The situation can, therefore, be tabulated as follows.

(B′)	c		v		s		u	
in labour	48	+	6	+	6	=	60	$\lambda_b = 5$
in money	24	+	3	+	3	=	30	$\rho_b = 2.5$
in quantity	9.6	+	1.2	+	1.2	=	12	$r = 0.11$

If this is the case in the spinning sector, however, it must be supposed that the worker in the sector which produces baskets of wage-goods too must work 12 hours a day, dividing it half and half between the necessary and surplus labour-time. It will be shown presently (section

III) that in this latter sector alone the v-column and the u-column must always remain proportional. Hence, if the same rate of profit of $r = 0.11$ as in the cotton spinning sector (B′) must prevail in the expanded wage-goods producing sector (A′) as well, the following situation must be obtained.

(A′)	c		v		s		u	
in labour	36	+	6	+	6	=	48	$\lambda_a = 3$
in money	18.6	+	3	+	2.4	=	24	$\rho_a = 1.5$
in quantity	12.4	+	2	+	1.6	=	16	$r = 0.11$

Are these two situations (A′) and (B′) mutually consistent? There must be something wrong since the values and prices are proportional ($\lambda_a : \rho_a = \lambda_b : \rho_b = 2 : 1$) even though the value composition of capital differs between the sectors (c/v is 6 in A′ and 8 in B′). Such a thing is, of course, impossible. Besides, the capital composition in (A′) is 6 in value and 6.2 in price (the latter being about 3.3% higher); but in (B′) it is 8 both in value and in price. This cries out for a justification.

3.3 WAGE GOODS SECTOR

Let us first establish the reason why only in the sector producing wage-goods the v-column and the u-column must be proportional. Let the number of workers engaged in the production of wage-goods be m, and let A be the total number of the baskets of wage-goods produced in society. Then we have:

$$(1) \qquad 48m = \lambda_a A, \; 24m = \rho_a A, \; 16m = A$$

necessarily.

Now let n stand for the total number of productive workers in society, L for the total number of hours worked in society, w for the wage-rate per hour, and e for the rate of surplus value. Then we have:

$$(2) \qquad 6n = L/(1+e), \; 3n = wL, \; 2n = A.$$

Under any circumstances, however, the identities

$$(3) \qquad \lambda_a A = L/(1+e)\,, \; \rho_a A = wL$$

must hold.[4] Hence in capitalist society as a whole the *v*-column times n/m should always be equal to the *u*-column in the production of wage-goods.

Now let us consider the question of capital composition in value and in price. Let only one capital-good X be in use in both sectors, and let X_a be the quantity of it in use in the wage-goods sector. If La is the number of hours of labour spent in that sector, the capital composition in value is $\lambda_x X_a(1+e)/L_a$ and in price $\rho_x X_a/wL_a$. Hence their ratio is $\lambda_x(1+e) : \rho_x/w$. In the spinning sector the same ratio applies inasmuch as the capital composition in value is $\lambda_x X_b(1+e)/L_b$ and that in price is $\rho_x X_b/wL_b$ there, with similarly defined symbols X_b and L_b.

In the case in which there are many capital-goods X, X′, X″ ... in use, it is not possible to expect the same ratio between sectors unless these capital-goods are used in the same fixed proportion everywhere. Though in each sector the ratio is in the form:

$$\text{Average of } (\lambda_x, \lambda'_x, \lambda''_x, \ldots)(1=e) : \qquad \text{Average of } (\rho_x, \rho'_x, \rho''_x, \ldots)/w,$$

the method of averaging (assigning weights to different capital-goods) differs from one sector to another. Even so, however, one can say the following. If the ratio $\lambda_x()(1+e) : \rho_x()/w$ is the largest with regard to the i-th capital-good and the smallest with regard to the j-th capital-good, then the ratio in any sector must be between these two extremes.

For the problem at hand, however, detail is not important. For simplicity it may be assumed that, if in the wage-goods sector the capital composition in price is $x\%$ higher than that in value, the same deviation applies in all other sectors, though in reality it varies within limits. In other words, let us assume either only one capital-good to be in use in all sectors or many capital goods in the same mix everywhere.[5]

3.4 A REVERSE SOLUTION

At this point one may ask whether it is possible at all to arbitrarily expand the spinning sector first, and then adapt to it the wage-goods sector, even though the latter has to satisfy the condition of proportionality between the *v*-column and the *u*-column. By a trial-and-error experiment one can easily discover that this is impossible. Perhaps Uno himself tried to construct a plausible example and could not obtain a satisfactory one. Actually, I do surmise that to have been the case, since he keeps the wage-goods sector rather vague without supplying necessary numbers. At any rate the reason that an extensive debate over Uno's demonstration of the law of value has taken place to no avail is due to the lack of appropriate numbers illustrating the condition of that sector.[6]

With this in mind I seek a solution by reversing the procedure, i.e., by fixing the conditions of the wage-goods sector as in (A') first, and then adapting the spinning sector to it as in (B") below, rather than doing the reverse. But as before I assume in (B") that the one-day output of a spinning worker is 12 kg. of cotton yarn, and that in these are embodied 60 hours of labour. Then the situation in that sector must be the following.

(B″)	c		v		s		u	
in labour	48	+	<u>6</u>	+	<u>6</u>	=	<u>60</u>	$\lambda_b = \underline{5}$
in money	24.8	+	<u>3</u>	+	3.1	=	30.9	$\rho_b = 2.575$
in quantity	9.63	+	1.17	+	1.2	=	<u>12</u>	$r = \underline{0.11}$

The numbers previously known are underlined.

Here 12 of the 60 hours of labour are current, so that 48 hours of labour must be embodied in the means of production. Since the necessary labour-time is 6 hours, the value composition of capital is 8. The price composition of capital must be about 3.3% higher, as in the wage-goods sector. It should then be about 8.267. This requires that the capitalist invests 24.8 shillings in the means of production per employment of one worker. Since he has to realise the general rate of profit of 11% on $c + v = 27.8$ shillings in order for the labour theory of value to hold, his profit must be 3.1 shillings. Hence the output, the 12 kg. of cotton yarn, must be sold for 30.9 shillings or for the unit price of $\rho_b + 2.575$.

In this case the two sectors (A′) and (B″) are consistent, and the equality:

"2 baskets of wage-goods = 1.17 kg of cotton yarn"

holds unambiguously. For both sides embody 6 hours of necessary labour and can be bought for 3 shillings, which constitute the daily wage.

3.5 NECESSARY AND SOCIALLY NECESSARY LABOUR

Here it is correct to say that 1.17 kg. of yarn are the product of the necessary labour of 6 hours, but not of *socially* necessary labour of 6 hours. Indeed, since the value of yarn is $\lambda_b + 5$, 1.17 kg. of yarn must embody only 5.85 hours of socially necessary labour. (It is, of course, part of elementary knowledge that "necessary labour" and "socially necessary labour" are two entirely different concepts.) The spinning capitalist, however, recovers the money value of variable capital advanced by selling 1.17 kg. of yarn for 3 shillings. The reason for this is that the worker can buy back 2 baskets of wage-goods for 3 shillings and these are necessary and sufficient for the reproduction of labour-power used up for the day. These 2 baskets of wage-goods are not only the product of the necessary labour of 6 hours, but they also embody 6 hours of *socially* necessary labour.

If this is the case the following famous statement of Uno does not appear to be accurate in this case.

> Suppose, for example, that the spinning capitalist sells 6 labour-hours' worth of his output for 3 shillings. If, despite this, the other capitalist sold only 5 labour-hours' worth of his output for 3 shillings to the worker, then not only would the latter fail to receive sufficient means of livelihood, but the capitalist who produces the means of livelihood would gain more than the capitalist who spins cotton, which would make the production of cotton an altogether futile enterprise for capital.[7]

This statement is correct in the absence of surplus labour (i.e., when $r = e = o$); and Uno meant it for such a case, for example, as between (A) and (B) above. Indeed while the spinning capitalist sells 1.2 kg. of cotton yarn which embody 6 hours of labour for 3 shillings, the capitalist who supplies wage-goods cannot sell only 1.67 baskets of them which embody 5 hours of labour for 3 shillings. For that would mean that the spinning worker cannot reproduce his labour-power. But that merely reasserts the fact that in the absence of surplus labour values and prices are proportional, and that products embody as much "*socially* necessary" labour as they do "necessary" labour.

In the more general case with positive surplus labour (r, e, > o) the equivalence of "necessary" and "socially necessary" labour fails. Although every worker performs 6 hours of necessary labour and is paid 3 shillings, the latter enable him to buy back the product of 6 hours of (socially necessary) labour only as wage-goods. In the case of cotton yarn for the production of which a higher value composition of capital than that in the wage-goods sector applies, the wage of 3 shillings enables him to buy back the product of only 5.58 hours of (socially necessary) labour. It enables him to buy back the product of more than 6 hours of (socially necessary) labour, if it is of the sector whose value composition is lower than that in the wage-goods sector.

Suppose that a capitalist produces, per employment of one worker, 10 units of (C) which embody 42 hours of labour . Then the value composition of capital (c/v) in this sector is 5 and is smaller than in the sector (A′) which produces wage-goods which is 6. The situation in that case is as follows:

(C)	c		v		s		u	
in labour	30	+	6	+	6	=	[42]	$\lambda_c = 4.2$
in money	15.5	+	3	+	2	=	20.5	$\rho_c = 2.05$
in quantity	7.56	+	1.46	+	0.98	=	[10]	$r = 0.11$

The composition of capital in price is 5.167 or about 3.3 % higher than that in value. In this case what the wage of 3 shillings can buy back is 1.46 units of (C) in which as much as 6.132 hours of (socially necessary) labour are embodied.

Yet all these industries (A′) (B″) and (C) operate at the uniform profit rate of 11%, so that it should not be, to any one in particular, "an altogether futile enterprise for capital".

3.6 THE NECESSITY OF THE LAW OF VALUE

What is the relevance of the above technicalities to the idea of the necessity of the law of value?

In capitalist society the production of commodities is universal and not partial. In other words, all use-values that society needs are produced as value-objects because even labour-power is converted into a commodity. Value-objects tend to embody only socially necessary labour. When all commodities embody only socially necessary labour, they are all produced in quantities that meet the social demand (none being either overproduced or underproduced). This further implies that the allocation of productive labour in society is optimal and a uniform rate of profit obtains in all spheres of production. The law of value, on this ground, claims that all commodities tend to be exchanged at equilibrium prices, equilibrium prices which presuppose an optimal social allocation of productive labour, and which consequently also presuppose the expenditure of socially necessary labour for the production of all commodities.

The direct producer in a capitalist society cannot live for an extended period of time unless he sells his labour-power at least for its value. If the consumption of his labour power per day can be made good by 2 baskets of wage-goods, as supposed above, the socially necessary labour for the production of these two baskets is also the necessary labour for the reproduction of his labour-power. Though every worker works for 12 hours, he receives the wage of 3 shillings with which to buy back 2 baskets of wage-goods produced with 6 hours of labour. As soon as this assumption is made, it is possible to deduce the price of any other commodity the price at which that commodity must be traded, if its daily output per employment of one worker is known in physical quantity and in terms of the labour directly and indirectly spent on it.

Now let us consider a "simple commodity producer" who produces 5 units of (D) which are the product of 40 hours of labour directly and indirectly. Assume that he too works 12 hours a day and can reproduce his daily consumption of his labour-power with 2 baskets of wage-goods, like the worker employed by a capitalist. In this case he can continue to sell his product for any price that enables him to recover the advance of his means of production and to pay for his upkeep without getting into any serious trouble. For example, a situation such as

(D)	c		v		s		u	
in labour	28	+	6	+	6	=	40	$\lambda_d = 8$
in money	14.46	+	3	+	0.88	=	18.34	$\rho_d = 3.668$
in quantity	3.94	+	0.82	+	0.24	=	5	$r = 0.05$

can be repeated endlessly. Here the profit-rate is only 5%. But that may be enough for the small producer to survive. The problem here is that under other circumstances he may also get a profit-rate of 10%, 15%, etc.; and the price of his product (ρ_d) varies accordingly. In any case it is not possible to determine a definite price corresponding to the socially necessary labour for the production of (D), unless the same commodity is capitalistically produced in the same market.

If, however, this commodity (D) is capitalistically produced, then the following situation must necessarily hold per employment of one worker per day:

(D′)	*c*		*v*		*s*		*u*	
in labour	28	+	6	+	6	=	40	$\lambda_d = 8$
in money	14.46	+	3	+	1.92	=	19.38	$\rho_d = 3.876$
in quantity	3.73	+	0.77	+	0.5	=	5	$r = 0.11$

In this case society's productive labour is allocated in such a way that in all branches of production, including the production of (D), the uniform profit-rate of 11% obtains.

The fact that the 5 units of (D) are sold for 19.38 shillings (i.e., $\rho_d + 3.876$) confirms the 40 hours of labour spent on them directly and indirectly as that which was socially necessary.

In small commodity production, however, a condition of excess supply can persist for a lengthy period of time in which the producers of (D) are obliged to sell its 5 units for only 18.34 shillings and to sustain the low profit-rate of 5%. This would in effect mean that the 8 hours of labour actually spent for the production of (D) per unit were not all socially necessary, and that only their 94.6% (= 18.34/19.38) or 7.57 hours were formative of value. In other words, value cannot be defined properly unless society's productive labour is (or tends to be) optimally allocated to the production of all commodities.

The above is my interpretation of what Uno meant by the "necessity of the law of value" in capitalist society. The exchange of commodities "according to values" does not mean "at prices proportional to values." It rather means the exchange of commodities "at prices which reflect an optimal allocation of society's productive labour to the production of all commodities."

Notes and References

1. Kozo Uno, *Principles of Political Economy*, translated from the Japanese by Thomas T. Sekine, Sussex: Harvester, 1980, p. 23.
2. Ibid., p. 25.
3. Ibid., p. 26.
4. The first relation states the fact that the working-class buys back all the wage-goods produced in society with the necessary labour-time that it performs. The second relation says that all the wage-goods produced in society are entirely purchased by the total wages-bill paid to the workers.

5. This condition may, at first sight, appear to be rather unduly restrictive. However, the same condition is implied by the usual formulation of the law of average profit that the prices of commodities with a higher(lower)-than-the-average value composition of capital are higher(lower) than their value-prices, i.e., prices proportional to values. See Thomas T. Sekine, *The Dialectic of Capital*, vol. II, Toshindo Press, Tokyo, 1986, pp. 115–23.
6. There is enormous literature in Japanese which is devoted to trying to figure out what exactly it was that Uno tried to establish in those enigmatic paragraphs of his *Principles* (1980).
7. Uno, *Principles*, p. 25.

4 The Demonstration of the Law of Value and the Uno–Sekine Approach

Sadao Ishibashi

4.1 INTRODUCTION

Since the publication of Marx's *Capital*, Volume III, controversies over the labour theory of value have, broadly speaking, taken place in two parts. One has dealt with the so-called transformation problem, and aims at a clarification of the proper manner in which production-prices may be derived from values. The other has concerned itself with the demonstration of the law of value as such. In this case the argument has centered on the manner in which the quantities of labour embodied in commodities regulate their prices. It has sought to determine the circumstances under which commodities can be exchanged at value, i.e., commodities which embody the same amount of labour may be exchanged at par for one another.

These two problems are closely related to each other. Those who address the first problem have tried to establish the validity of the labour theory of value indirectly, i.e., by showing that production-prices could indeed be rigorously *derived from* values. In this approach, they do not seek further verification of the law of value itself. For if production-prices hold true, the values from which they are derived must also hold true. However, the validity of Marx's labour theory of value itself has always been questioned (for instance, by Böhm-Bawerk). The second problem (i.e., why production-prices must be derived from labour values, and not from anything else), therefore, consists of defending the validity of the labour theory of value itself, with only indirect reference to the question of the transformation.

In the present paper we are chiefly concerned with the second problem, but we must keep in mind a precise understanding of the transformation problem as well. For the law of value which we shall demonstrate as valid is consistent with production-prices which diverge from value-proportionality.

As is well known, Marx's discussion in Volumes I and II of *Capital* is predicated on the assumption that commodities are bought and sold at prices proportional to values. Only in Volume III does he develop production-prices which diverge from value-proportionality. Thus we have two kinds of prices: **production-prices** that bring the market into equilibrium and **value-proportional prices**. The relation between the two cries out for clear explanation.

Several explanations have so far been offered. One is the logical-historical theory according to which value-proportional prices did indeed hold true in a society of simple commodity producers. This idea originates with Marx himself, because he tried to demonstrate the labour theory of value in chapter 1, of *Capital*, Volume I, on "Commodities". Kozo Uno rejected this logical-historical approach to the law of value. He held the view that the law of value constitutes the most fundamental law of capitalism, and that its demonstration crucially depends on the conversion of labour-power into a commodity. From this point of view it is clear that the demonstration of the law of value cannot be undertaken except in the context of the production-process of capital. To try to establish the law in the context of simple commodity circulation would, according to Uno, be asking the impossible.

Many Marxian theorists in Japan sympathised with Uno's approach but were not fully convinced, since Uno's own demonstration was imperfect. Thus an extended controversy over the "demonstration problem" ensued. The most widely held conclusion of the debate during the 1960s was that "exchange of commodities at value" in the sense of the exchange of commodities embodying an equal quantity of labour at par cannot be demonstrated, nor is such a demonstration theoretically called for. I will refer to it as the Suzuki-group conclusion, since it was shared by the followers of Suzuki's main opus *Keizaigaku-Genriron* (Kohichiro Suzuki, 1960, 1962).

In the 1970s a reassessment of that position began to appear, which suspected that "the exchange of equal labour time" had been dismissed too lightly. Had not the Suzuki-group thrown away the baby with the bathwater? It was thought that the "exchange of commodities at value" in the sense of the "exchange of commodities embodying an equal amount of labour time" merited further investigation. This view continues to be held by many Marxian economists to this day, even though the initial enthusiasm over that debate subsided during the 1980s.

One of the purposes of this essay is to examine, in light of the Japanese debate, Sekine's two papers on value theory appearing in this volume. By doing so, I wish to bring out the merits and demerits of Sekine's work, which was undertaken outside Japan and hence has so far not been viewed in the context of the controversy over the demonstration problem in Japan. I hope to conclude this paper by presenting my view on the matter, after a due survey of the contributions by other authors.

4.2 UNO'S VIEW ON THE LABOUR THEORY OF VALUE

Uno's main ideas are stated in his *Principles* (1980). He criticises Marx for referring to the substance and magnitude of value in Simple Circulation. His contention is two-fold. First, Marx's premature reference to value substance muddled his (Marx's) formulation of simple circulation-forms such as commodity, money and capital, and failed to bring out the full significance of the study of these forms. Secondly, Marx's "demonstration" of the law of value contains serious logical flaws.

Uno, therefore, departs from the three volumes of *Capital*, and rearranges the same theoretical content in a different order in his *Principles*. The latter is divided into the three doctrines of circulation, production, and distribution. In the doctrine of circulation, he endeavours to develop circulation-forms only, without direct reference to value-creating labour as "social substance". He claims that such circulation-forms as commodity, money and capital must be examined independently of (that is without making explicit reference to) the production-process of capital.

On the other hand, he begins the doctrine of production with the concept of labour-and-production process prior to its being subsumed by capital, the last major category of the simple forms of circulation. He states that labour, as constituting the substance of value, can be elaborated only in that context after the subsumption of the labour-and-production process under the form of capital. Uno claims that he is, in fact, following Marx's own method in his novel attempt at demonstrating the law of value within the doctrine of production. I consider this part of the *Principles* to be the touchstone of Uno's theoretical contribution.

How does Uno approach the problem? In a nutshell, I believe that it may be summed up in the following statement. "Marx's argument lacks a positive proof of the labour theory of value in terms of capitalist-necessity" (Uno, 1980, 32). According to Uno, Marx merely extracts "one and the same sort of labour, human labour in the abstract", from the exchange relationship of two commodities, i.e., from the exchange equation "1 quarter corn = *x* cwt. iron", as something common to both sides and which may be viewed as value, once one has distilled away their use-value properties (Marx, 1971a, 45–6). Marx believes that the magnitude of the commodity's value must then be determined by the labour time socially and *technically* necessary for its production.

Uno (1980, 32–33) first argues that "abstraction from use-values" occurs objectively in the production-process of capital, in which the capitalist can actually choose to produce any use-value of his choice. It does not occur, except subjectively, if we merely observe on-going commodity exchanges from the outside. Secondly, Uno argues, "the average labour-power of society" which is supposed to produce value can become a reality only in the production-process of capital in which labour-power purchased as a commodity performs indifferent labour. The de-skilling of labour, etc., which value-forming labour presupposes, becomes real and objective also in this condition, and not merely in our imagination as we observe the exchange of commodities from the outside. For these reasons, Uno establishes the proposition that "labour forms the substance of value" in the context of the production-process *of capital*, not in the context of commodity production in general.

Two concepts are crucial in Uno's demonstration of the labour theory of value. One is that of the labour-and-production process and the other is the idea that "the worker buys back the product of his necessary labour".

Uno's labour-and-production process is quite different from Marx's "labour-process", though they can be easily confused because of their similar names. To Uno the labour-and-production process is the logical ground for the explication of the substance of value, that is to say, it constitutes the substantive base of the economy. For when it is subsumed under capital, it becomes the production-process of capital. It must contain all the ingredients of use-value production, common to all societies, as the general norms of economic life.[1] The capitalist economy enfolds these general norms within its inner laws or structures.

More specifically, these ingredients include (1) the dual property of productive labour, concrete-useful and abstract-human, referred to here for the first time. Uno (1980, 32) emphasises that these two aspects of productive labour, one qualitative and the other quantitative, are common to all societies, and not specific to commodity production. In capitalism, abstract-human labour **becomes** value-forming labour. They also include (2) the division of the working-day into necessary labour time and surplus labour time. This division too is common to all societies. As Marx (1971a, 226) says, "capital has not invented surplus labour". It is important to recognise that, according to Uno, surplus labour can exist even in a classless society, although, needless to say, surplus labour produces surplus *value* only in the value-augmentation process of capital.

Let us now examine Uno's idea that the worker "buys back" his own product (1980, 26). Sekine's articles contain numerical illustrations of this concept. Hence I will limit myself to commenting only on Uno's main theoretical ideas behind the numbers.

First, in approaching the demonstration problem, Uno assumes that the worker performs only necessary (and not surplus) labour, i.e., he assumes the process of value formation (exclusive of value augmentation). Secondly, as conclusions from the demonstration problem, Uno tries not only to attribute the substance of value to labour, but also to establish that the "exchange of commodities at value" amounts to an "exchange of commodities embodying an equal quantity of labour time". Thirdly, Uno regards it as the pivot of his argument that the worker has to buy back the products of his own labour as commodities with the money wages that he receives in return.

In the labour-and-production process, common to all societies, human beings can obtain use-values only by working on nature, either directly or indirectly, and in the process consuming both their labour-power and means of production.

In the production-process of capital exactly the same thing is supposed to occur though mediated by the commodity-form. That is to say, the capitalist who is a money-owner purchases both labour-power and means of production as commodities and obtains the product, also as a commodity. This he can do by letting his workers expend their productive labour with the aid of an appropriate mix of means of production which serves his purpose. The direct producers are paid money wages, but do not have direct access to the product of their labour. They are consequently obliged to "buy back" their means of livelihood, the product of their own labour, with their wages.

This is the quintessence of the "production of commodities by means of commodities" (Uno, 1980, 25), or the labour-and-production process operated under the commodity-economic forms. In order for this process to repeat itself, however, prices at which commodities are exchanged among themselves must be regulated in a logical way. For otherwise the general norms of economic life may fail to be satisfied, which would paralyse economic life itself. Above all the prices must be such as to ensure the replacement of labour-power and means of production currently consumed.

Obviously, in the present case (without surplus labour), such prices are proportional to values so long as the money wage-rate per hour is the same in all fields of production.[2] For otherwise some capitalists can realise a surplus and others a deficit, after replacing whatever labour-power and means of production they currently consume. This point is demonstrated in Sekine's numerical examples.

Thus, in this case, Uno succeeds in demonstrating that "labour forms values" by pointing to the necessity that commodities must be exchanged according to the labour time embodied in them. Thus, according to Uno, the two general norms (1) that the provision of use-values requires an expenditure of human labour, and (2) that the reproduction of economic life presupposes (implicit or explicit) exchanges of products both appear in the commodity-economy in the form of the law of value, which regulates the terms of exchange of commodities.

Unfortunately, Uno's demonstration of the labour theory of value is satisfactory only in the above case, i.e., in the absence of surplus labour. Once the workers perform surplus labour, prices of commodities diverge from proportionality to embodied labour time in commodities. Equilibrium prices (production-prices) deviate from value-proportional prices, if the organic composition and the turnover-time of capital differ from one industry to another. Uno was fully aware of this point. Yet he continued to assume value-proportional prices throughout the doctrine of production, regardless of whether surplus labour is performed or not. What should we make of this?

4.3 THE DEBATE OVER UNO'S DEMONSTRATION

Uno succeeded in improving the method of demonstrating the law of value, but what he demonstrated in the above case was the same as Marx's claim, namely, that commodities embodying an equal quantity of labour are to be exchanged for one another at par. If so, Uno should have taken one step forward in explaining the relationship between labour values and production-prices. For in a real (as opposed to hypothetical) capitalist economy, commodities are not exchanged at their values.

The Suzuki-group economists excoriated Marx's conversion of values into production-prices.[3] They also denied the validity of Uno's demonstration of the law of value in consequence. That is to say, the group claimed that Marx's theory of production-prices requires a modification of the idea that commodities produced with an equal quantity of labour should be exchanged for one another at par. Indeed Marx's method consists of showing first that different profit-rates would arise if commodities were exchanged at values, and thus he introduces capitalist competition to eliminate these differences and equalise profit-rates (Kohichino Suzuki, 1962, 492–504).

There are two flaws in Marx's transformation method. One is that it skips the conversion of the cost-price of a commodity from value terms to price terms. The other is that capitalist competition, which performs the transformation, might bring about a reallocation of society's productive labour. These two problems must be addressed by all theories of transformation from values into prices.

The Suzuki-group insists on the following two points. First, it emphasises that there is a dimensional difference between the first two volumes and the last of *Capital*. (In other words, values and prices belong to different dimensions.) Secondly, the proposition that commodities are exchanged at values should be dropped altogether from the content of the law of value.

The first point implies that value is a substantive concept whereas price is a formal concept. Substance cannot be converted into forms. Production-prices are the forms of enforcement of the law of value. Values, so it is claimed, cannot be transformed into production-prices at all. From this point of view, the Suzuki-group economists deny the very existence of the transformation problem itself.

This argument is based on the claim that the dimension of "capital in general" involves no competition, and must be sharply distinguished from the dimension of "competition among individual capitals". The doctrine of production, in other words, makes no reference to the relation of exchange among individual capitals. If so, it follows that the scope of the law of value, as the inner law of capitalism, should be restricted to the stipulation of labour as constituting the substance of value. The theory of surplus value must then be formulated as a relationship between capital as a whole and wage-labour as a whole.

This claim leads to the second point of the Suzuki group that the exchange of commodities at values is an absurd proposition. This group, therefore, does not recognise the thesis that production-prices must somehow be regulated by values as a valid theoretical problem.

From such a point of view, many questions were raised over Uno's demonstration of the law of value. While accepting Uno's thesis that the law of value must be demonstrated in the doctrine of production, rather than that of circulation, many economists failed to appreciate the significance and validity of his "demonstration." The absence of a convincing explanation as to why, even in the presence of surplus labour, Uno followed Marx in assuming exchanges of commodities at value-proportional prices was viewed as a fatal flaw.[4]

During the 1970's, however, some criticisms of the Suzuki-group thesis began to appear. Roughly speaking, two points of view were put forward. One stated that, even in the dimension of the doctrine of production, individual capitals must not be ignored, and that the relation between capital as a whole and wage-labour as a whole can be explained in commodity-economic terms only in light of competition among individual capitals (Mawatari, 1970). The other point of view insisted that production-prices should not be disconnected from the value-creating labour substance. This latter claim implies that production-prices and values belong to the same dimension, since production-prices as equilibrium prices are not as indeterminate as mere market-clearing prices which solely depend on conditions of demand and supply. In either case, it was felt that a strict application of the Suzuki-group approach might render the concept of value empty. The critics of the Suzuki school can be divided into three groups.

The first group concurs with the Suzuki school in its insistence that no reference should be made to the exchange of commodities at values, but it differs in that it reaffirms the validity of the transformation problem. This group tries to clarify how the substance of value regulates prices which are not proportional to values. Itoh, who belongs to this group, makes the following claim: "Though the substance of value ($c_i + v_i + s_i$) of a commodity does not proportionally regulate the standard price or the form of value, its portion ($c_i + v_i$) must thus work as a substantial core regulator of the standard price in the mutual relations of commodities. . ." He also states, "Their [standard prices'] deviation from prices proportional to the labour substance must, however, be restricted within the range of the redistribution of surplus labour across industries" (Itoh, 1988, 134). We can easily see that he is implicitly making reference to Uno's demonstration.[5]

The second group is in no way inhibited from referring to the exchange of commodities at values, even in the presence of surplus labour. For it takes the situation of "no surplus value" as a simplified model. Its approach appears to be similar to Uno's, except for its explicit supposition that the value composition of capital should be the same in all industries, and its disavowal of the proposition that the exchange of commodities at equal values can constitute the core of the labour theory of value (Yamaguchi, 1985, 131). Value-proportional prices are not deemed real but only notional. Yet the fact that prices are proportional to values in the absence of surplus labour is interpreted to mean that fluctuating market prices gravitate towards the levels dictated by the social allocation of labour. Prices diverge from values because capitalists can flexibly give and take the surplus-value component of the commodity values (Yamaguchi, 1985, 125–7).

The third group regards the exchange of commodities at values as an essential component of the law of value, and believes that such an exchange should be explained in the doctrine of production even by the artifice of envisaging a special type of competition which is different from that which actually prevails among capitals for a higher rate of profit. What sort of competition is it? We can list three different kinds of notional competition that have been invented. The three different types of capitalist competition, which I describe in the following, were all believed to yield value-proportional prices instead of production-prices in the market.

In the first case, individual capitalists would pursue prices which are proportional to the value of each commodity, so that $q_i/(c_i + v_i + s_i)$ would be the same for all i (Mawatari, 1979b, 142). In the second, capitalists would compete for the highest profit per surplus labour embodied in the commodity π/s, instead of the highest profit-rate: π/κ (Kobayashi, 1977, 118).[6] In the third case, capitalists would compete for the highest ratio of profit to value added: $\pi/(v + s)$ (Miwa, 1988, 16–22).[7] The latter two cases are basically the same. For π/s is equal to $\pi/(e \times v)$, and $\pi/(v + s)$ is equal to $\pi/[(1 + e) \times v]$, so that these two types of competition have to suppose equalisation of the rate of surplus value. (The symbols used in this paragraph are defined as follows: π = selling price – cost-price; $e = s/v$; $\kappa = c + v$.) In the final analysis, these two cases seem to be the same as the second group above insofar as the portion of (c) is excluded from consideration as a factor that influences competition, which also amounts to ignoring differences in the organic composition of capital.

I myself once adopted a position similar to the third (Ishibashi, 1992, 93–126). The conclusion, however, depends on several assumptions, the arbitrariness of which can be legitimately criticised. In fact, the only rational competition in the capitalist market is the one in which the highest rate of profit: π/κ is pursued. It may, therefore, be wholly unwarranted to seek to demonstrate the possibility of commodity exchanges at values in the doctrine of production. The question of "demonstration" really boils down to whether or not the exchange of commodities at value is a useful device in order to make the labour theory of value more explanatory or convincing. Before reaching a conclusion on this point, let me examine Sekine's argument in some detail.

4.4 SEKINE'S CONTRIBUTION TO UNO'S THEORY

In summarising the two articles of Sekine, I consider "Seminar" to be the main source of information, and "Necessity" to be a supplementary one.[8]

"Seminar" is in three parts. The first part deals with the concept of value. The second covers the law of value in a concise fashion. This part, which can be supplemented by "Necessity" for details, contains arguments most relevant to the present discussion. The third part tackles the transformation problem and related issues. Following that structure, I shall summarize Sekine's argument in three steps.

(I) He begins by strictly distinguishing "value" from "exchange value" by which he means price or value-form. According to Sekine, value is "the property of a commodity which makes it qualitatively the same as, and only quantitatively different from, another commodity" (ch. 2, 19). On the other hand, the price of a commodity is nothing but an expression of its value in terms of a given quantity of the use-value of another commodity.

Value, however, need not be expressed proportionally by prices. The equation of exchange "1 quarter corn = x cwt. iron" only shows that both sides are equally priced in trade, and does not mean that they contain an equal magnitude of value. By his distinction of "the labour theory of value" from "the labour theory of price", Sekine suggests that the law of value cannot imply a theory of exchange of commodities embodying an equal amount of labour time (ch. 2, 20). Why then can we call prices, which are not proportional to values, the forms of value?

We can, of course, say that commodities are socially homogeneous as value. It goes without saying that this "homogeneity" is shared by all commodities which are offered by their owners in exchange for other commodities which they desire. Therefore, the expression of value in a price (value-form) by the owner of a commodity represents his desire to exchange it for another commodity.

As Sekine mentions, what value-form theory does is to explain how the equivalent commodity or "little money" which first arises in the simple value-form divests itself of all qualitative and quantitative restrictions, until it becomes the general equivalent or full-fledged money which may take the form, for instance, of gold. The latter then is the universal equivalent in the sense that it reflects the value of all commodities (ch. 2, 23). The emergence of the general equivalent or money opens up the possibility of indirect trade. Thus commodities can be sold for money rather than being directly exchanged for some specific use-value.

The ultimate expression of value takes the form of a money price, but it is, even then, nothing more than a subjective valuation of the commodity by its owner. The supply price must, therefore, be subject to revision in response to the degree of intensity of demand. Value can be measured objectively only when the commodity is repetitively purchased by money. Although the price fluctuates in response to demand and supply, Sekine believes that it eventually settles to a normal level. What he calls the "normal price" at this point becomes the equilibrium or production-price later, as the real parameters necessary for its determination are brought out explicitly (ch. 2, 25).

Why can he suppose the existence of stable normal prices? The reason is that behind normal prices he implicitly presupposes an equilibrium allocation of society's productive resources. Thus, if a commodity is neither over-produced nor underproduced relative to the existing pattern of social demand, so that the productive resources of society are properly allocated to all spheres of commodity production, normal prices which reflect values will be established. Sekine, however, makes no assertion that commodity values are proportional to normal prices (ch. 2, 26). He openly admits that normal prices are in general not value-proportional. Does this constitute a contradiction?

Sekine denies the existence of any contradiction on this point. Normal prices tell us that the commodities have been produced in the right quantities (as use-values) relative to the social demand. Their values imply that their production has, in each case, consumed the right quantities of society's productive resources (ch. 2, 26). It is true that normal prices need not be proportional to values under these definitions. Can we, however, claim that normal prices *express* values? At any rate, it is quite clear that Sekine does not want his law of value to have anything to do with the exchange of commodities at labour values.

So far Sekine has used the term "society's productive resources" without explicitly stating what constitutes the substance of value. But, to him, society's productive resources are represented by productive labour, the expenditure of which constitutes the substance of value. He defends that proposition as follows.

Since value is a qualitatively uniform property of a commodity, it is entirely justified to ask where that uniformity or homogeneousness comes from. Capital can produce any commodity if it is profitable to do so because of the conversion of labour-power into a commodity. Capital is also freely mobile between different spheres of production. This, according to Sekine, "amounts to saying that capital produces commodities as value, i.e., with indifference to their use-values" (ch. 2, 26). He regards the abstract-human labour, which labour-power purchased by capital expends, as forming the homogeneous substance of value. With regard to other factors of production, Sekine denies their capacity to form any homogeneous substance. According to him: "there are no such things as abstract-spatial land and abstract-physical capital"(ch. 2, 27).

What about the quantitative side of value? It is, of course, the quantity of value-forming labour (its abstract-human quality) that also determines the magnitude of value. Yet this fact does not imply that the labour time that is technically required for production *ipso facto* creates value. Sekine emphasises that "socially necessary labour forms the substance of commodity values" (ch. 2, 28). Here socially necessary labour is labour that is technically required to produce a socially necessary (equilibrium) quantity of the commodity. Thus when value is formed with socially necessary labour, the commodity is also supplied in the right (socially necessary) quantity, and its price must be "normal".[9] It is in that sense that Sekine claims that a normal price expresses the value of the commodity. He further claims that "socially necessary labour is the only real cost to society in the production of commodities" (ch. 2, 28).

(II) We shall now proceed to the main topic of the present paper, the necessity of the law of value. Sekine believes that "an extensive debate over Uno's demonstration of the law of value has taken place to no avail" because in his numerical illustration, Uno "keeps the wage-goods sector rather vague without supplying necessary numbers" (ch. 3, 56). From this point of view, Sekine tries to improve upon Uno's demonstration by supplying the "enigmatic paragraphs" of the *Principles* with a set of additional numbers which he believes are necessary.

In various steps Sekine confirms the thesis that only in the absence of surplus labour are values and prices proportional. In that special case alone, "products embody as much socially necessary labour as necessary labour" (ch. 2, 27). In the more general case with positive surplus labour, prices are not proportional to values, owing to interindustry differences in the organic composition of capital. These conclusions are widely known. The contentious point is how to interpret the "demonstration problem" in light of these commonly shared conclusions.

Sekine's view, consistent with his understanding of the value concept, is the following: "The law of value . . . claims that all commodities tend to be exchanged at equilibrium prices, equilibrium prices which presuppose an optimal social allocation of productive labour, and which consequently also presuppose the expenditure of socially necessary labour for the production of all commodities" (ch. 2, 27). Thus "the exchange of commodities according to values does not mean at prices proportional to values" (ch. 2, 30), but rather at normal prices, or to put it more concretely, at production-prices. For only such prices can ensure both "the validity of the labour theory of value and the viability of capitalist society" (ch. 2, 30).

I would like to locate Sekine's theory in the context of the Japanese debate on Uno's demonstration. Sekine's labour theory of value implies that socially necessary labour forms the substance of value. Hence the law of value sees to it that commodities are exchanged at normal prices, which reflect an optimal allocation of society's productive labour. He criticizes the idea that the law of value should be demonstrated as the exchange of commodities which are produced with an equal amount of labour time. In that sense, Sekine may be construed to belong to the Suzuki-group. But there is a very fundamental difference as well.

Even though Sekine does not believe that commodities are exchanged literally at value-proportional prices in the doctrine of production, he completely rejects the idea that the doctrine of production has to do only with "capital in general". That is why he, unlike those belonging to the Suzuki school, takes the problem of transforming values into prices quite seriously.

(III) In the third part of "Seminar" Sekine outlines his views on the transformation problem. The space does not permit me to go into detail. So I intend only to touch upon some of his main ideas.

Sekine first specifies what he calls "the technology complex", or a tabulation of all technical parameters needed for the determination of equilibrium. This complex must satisfy the condition of self-replacement. His value system based on it enables him to determine all values and the rate of surplus value. The value system is bound by the constraint that the workers should receive the product of their necessary labour. Sekine also has a price system which, in conjunction with the technology complex, can be solved for all prices and the rate of profit. This price system is subject to what he calls "the fundamental constraint of the capitalist market", which states that the total money value of the wage-goods should be equal to the total wages bill paid to the workers (31).

Sekine distinguishes between the dialectical (or conceptual) transformation and the mathematical (or quantitative) transformation. In the former case, the transformation is one-way, i.e., from values into prices. In the latter case, the transformation goes both ways, i.e., from values into prices as well as from prices into values. He has numerical examples to illustrate his theory (22–5).

His conclusion is the following. If we know the values and the rate of surplus value in advance, we can calculate the prices of the means of production and the rate of profit without directly solving the price system. Conversely, if we know the prices and the rate of profit, we can obtain the values of the means of production and the rate of surplus value without directly solving the value system. Values and prices are the two different faces of the same technology complex (35).

Of course, it turns out that they are in general not proportional to each other. For all that, "the deviation of equilibrium prices from values is by no means arbitrary. . . . Equilibrium prices are, as it were, tethered to values" (35–6). He claims that the law of average profit defines "the concrete mode of enforcement of the law of value through the motion of prices in the capitalist market" (36). The law of average profit, according to Sekine, is in two parts. The first law of average profit which shows the extent to which prices diverge from proportionality to value is explained by inter-industrial differences in the value composition of capital. The second law of average profit shows that prices diverge further from values as the rate of surplus value rises. For, if the rate of surplus value were zero, there would be no divergence (36–8).

Thus by insisting that equilibrium prices are ultimately tethered to values or embodiments of socially necessary labour, Sekine differs distinctly from the Suzuki group. Three propositions seem to play crucial roles in Sekine's argument. (1) In the absence of surplus labour, prices are bound to be value-proportional. (2) The extent to which prices diverge from value-proportionality depends on the extent to which the organic composition of capital in each case departs from its social average. (3) Competition among individual capitals for surplus profits results in achieving the law of average profit which enforces the law of value.

4.5 CONCLUSION

According to Sekine, socially necessary labour is "the only real cost" to society. Since socially necessary labour forms the substance of value, it follows that the value of a commodity represents the real cost that society pays in acquiring it as use-value from nature. If so, it seems only natural to conclude that the real cost of a commodity expresses itself in a definite quantity of another commodity for the production of which the same real cost has been paid to nature.

In the hypothetical case of no surplus value, it is easy for us to understand that values regulate prices. This connection becomes less obvious when surplus labour is performed, as prices diverge from value-proportionality. Even in that case, however, prices are "tethered to values" and, as Sekine points out, the deviation of prices from values is by no means arbitrary. Can we then declare, with Sekine, that values do not cease to regulate prices even in the presence of surplus labour? I believe we can.

But why, in the presence of surplus labour, cannot values be expressed by value-proportional prices? The reason simply is that the surplus-value component of the commodity value has a certain measure of flexibility in its replacement as part of continuing reproduction of the commodity (Itoh, 1988, 133–6). In other words, it is not necessary that the total amount of surplus labour embodied in each commodity should accrue to the capitalist who produced it. Surplus value produced in society can be shared by all capitals according to the magnitude of their advance of capital. Production-prices which result from the competition of capitals for a higher rate of profit function as the distributor of surplus value to individual units of capital.

What is the significance of this distribution if we look at it in light of the general norms of economic life? Such a question has never been raised, but is worth investigating.

It has been maintained that there is "fundamental freedom in dealing with surplus labour" in the labour-and-production process (Itoh, 1988, 126). Such freedom, however, does not seem to be present in capitalist society. For the distribution of surplus value as average profit is enforced by competition among capitals. Can all societies other than capitalist deal with surplus labour freely?

If we recall the fundamental nature of capital which pursues the maximum rate of accumulation, we can see the following. The law of average profit which is achieved by competition simply brings out the fact that all units of capital can accumulate at the same speed. If a capitalist enterprise fails to earn an average profit, it clearly cannot expand as rapidly as the economy as a whole. If different spheres of production expand at different rates, the capitalist economy as a whole cannot be on the path of maximal expansion.

It is true that, in any class society, surplus labour is to some extent used arbitrarily to fit the needs of the ruling class. Surplus labour, however, is not performed entirely for the support and maintenance of the ruling class. According to Marx, even in a classless society "a definite quantity of surplus labour is required as insurance against accidents and by the necessary and progressive expansion of the process of reproduction" (Marx, 1971b, 819). "Insurance" and "progressive expansion" are, therefore, the two purposes for which surplus products are required by the process of reproduction, apart from the individual consumption of the ruling class.

These products cannot be disposed of arbitrarily in any society. Surplus labour that produces them must be distributed to different spheres of production in proportion to their stock of the means of production and livelihood, which in capitalist society are called constant and variable capital. That portion of surplus labour that constitutes the fund for accumulation must be shared by all parts of society in such a way as to guarantee a state of balanced growth in production. I would call that state "reproduction in equilibrium".

From the above observation we may conclude that all societies must satisfy the general norm of economic life in the allocation of surplus labour for accumulation. We should, therefore, legitimately ask the question: how is the general norm pertaining to "reproduction in equilibrium" achieved in capitalist society? It is, of course, through the exchange of commodities, i.e, through the system of prices that this general norm can be enforced.

Since capitalist society is a class society, a portion of surplus labour must be applied for the individual consumption of the ruling class. It is for this reason that the general norm which pertains to the distribution of total surplus labour to different spheres of production has been overlooked. However, if we ignore that part of surplus labour which produces consumption-goods for capitalists, the remainder must be so allocated to each sphere of production in such a way as to achieve "reproduction in equilibrium", i.e., a process of reproduction which grows at a maximal speed.[10] The competition among capitals for a maximum rate of profit is, in fact, the capitalist way of achieving this general norm of economic life.

The prices of capitalistically produced commodities must, therefore, be regulated in such a way as to ensure balanced growth of all capital units at a maximal speed. Now we may divide the value of commodities (c + v + s) into two parts: so-called cost-price (c + v) which consists of constant capital (c) and variable capital (v); and surplus value (s). These two parts differ in the manner in which they relate to the regulation of the price.

As already repeated, the total amount of labour embodied in cost-price must be replaced exactly even in the case of simple reproduction. That is to say, the amount of labour embodied in this part of commodity value ought to regulate prices in such a way as to ensure "the exchange of commodities at value" in the sense of "the exchange of commodities embodying equal labour time." Whereas the surplus-value portion (s), the substance of which is surplus labour, must be allocated among capitals in order to enable them to expand in a balanced fashion. The prices which express surplus value as profit, therefore, are regulated so as to achieve the allocation of surplus value in proportion to the magnitude of capital advanced. We can state that much from the point of view of the general norm.

Yet it must be noticed that the distribution of surplus value in proportion to capital advanced (here represented by cost-price) assumes that cost-price itself does not involve a transformation problem.[11] This ignores the fact that the prices of both constant and variable capital may not be proportional to their respective values in the presence of surplus labour. Nevertheless this fact does not repudiate the regulation of prices by the quantity of labour embodied.

It is in order to recover the same amount of labour for both constant and variable capital that prices must deviate from proportionality to values when surplus value is socially shared. Means of production and means of consumption spent for production of commodities must be replaced by new means of production and new means of consumption embodying in each case the same quantity of labour as before. That is nothing other than "the exchange of commodities at value" in the sense of "the exchange of an equal quantity of labour embodied therein." Such replacement is needed even for simple reproduction. On the other hand, surplus value must regulate prices in such a way as to distribute itself in a uniform ratio to the magnitude of capital advanced. These prices are production-prices in terms of money. Even though these prices are not proportional to values, they are still so regulated as to ensure the exchange of labour quantities consistent with the process of reproduction which expands at a maximal speed. For further details, see the Appendix.

The above is the way in which values regulate prices. In this sense we may claim that prices can express values. Needless to say, such normal prices can be realized concretely as production-prices together with the formation of an average profit in competition among capitals. In the doctrine of production, the law of value must be understood to the extent that it clarifies the significance of production-prices in view of the general norm of economic life.

Thus the law of value first of all implies that socially necessary labour forms the substance of commodity values, and secondly that such values should regulate prices so as to enable all units of capital to expand themselves at a maximal speed. This law of value expresses itself as production-prices.

Values, the substance of which is socially necessary labour, do not regulate prices as value-proportional prices, but as prices that make "reproduction in equilibrium" possible. The validity of this idea depends on the proposition that equilibrium under capitalism represents the supra-historic state of "reproduction in equilibrium". Prices are not only "tethered to" values, as Sekine mentions, but are also "regulated by" values. I hope to have clarified the meaning of what Sekine calls the problem of mathematical transformation from the point of view of the general norm. It explains the reason why labour time embodied in capitalistically produced commodities should regulate prices.

Appendix

Itoh (1988, 220–6) explains what in my interpretation constitutes the general norms in the context of the theory of production-prices instead of treating it within the doctrine of production. I wish to supplement my argument in the text over how the substance of labour regulates prices, by making use of Itoh's three tables (1988, 222), which I reproduce on the following page.

Itoh explains these tables as follows. Table 1 shows the labour-substance incorporated in the commodity product of each department of production, whereas Table 2 exhibits a set of production-prices which are formed on the basis of the substance of value embodied in the commodity products as listed in table 1. Table 3 displays the substance of value acquired by each department by selling its product at the production-prices as in table 2.

Controversies over the transformation problem since Bortkiewicz (Sweezy, 1949) have already stated production-prices which are shown in Table 2. These prices take into consideration the simultaneous decision on, and the interdependent relation between, the distribution of surplus value as average profit, and the conversion of the cost-price of the commodity from value terms to price terms.

Both constant capital (c_i) and valiable capital (v_i) in Table 1 are replaced as c_i (means of production) and v_i (means of consumption) in Table 3, which contain the same amount of labour-substance as in Table 1, through the prices of $c_i x$ and $v_i y$ which are not proportional to the amount of labour-time of c_i and v_i. On the other hand, surplus value (s_i) is distributed in such a way as to enable each department of production to expand itself on maximal balanced scales. Therefore, it should be noted that the substance of surplus value acquired (s_i') must be distributed in proportion to cost-price ($c_i x + v_i y$ in terms of money or prices of production), but not to $c_i + v_i$ (in terms of labour), because reproduction must be performed by exchange of commodities through prices.

Table 1 The Substance of Value Produced (a_i) (millions of hours)

Department of production	*Constant capital* c_i	*Valiable capital* v_i	*Surplus-Value* s_i	*Values of products* a_i
I	225	90	60	375
II	100	120	80	300
III	50	90	60	200
Totals	375	300	200	875

$s_i/v_i = 2/3$, $a_i = c_i + v_i + s_i$. From the numbers in this table the general rate of profit r is determined to be 25 per cent. Assume $z = 1/2$, then $x = 16/25$, $y = 8/15$.

Table 2 The Prices of Production (P_i) (millions of dollars)

Department of production	$c_i x$	$v_i y$	p_i	P_i
I	144	48	48	240
II	64	64	32	160
III	32	48	20	100
Totals	240	160	100	500

$p_i = r\,(c_i x + v_i y)$, represents the average profits. $P_i = (c_i x + v_i y) + p_i$, the prices of production.
$c_i x$ and viy represent constant and variable capital of each department in terms of prices of production. The sum of them $(c_i x + v_i y)$ is, therefore, cost-prices.

Table 3 The Substance of Value Acquired (a'i) (millions of hours)

Department of production	c_i	v_i	s'_i	a'_i
I	225	90	96	411
II	100	120	64	284
III	50	90	40	180
Totals	375	300	200	875

$s'_i = P_i \div 1/z = \sum s_i \times p_i / \sum p_i$, $a'_i = c_i + v_i + s'_i$

Notes and References

1. Uno distinguishes rigorously between the substance and the forms of the capitalist economy. The general norms of economic life consist of the rules which the substance of economic life must always abide by. They may also be regarded as features of economic life common to all societies and not peculiar to the historical forms of the commodity-economy. We must, however, take note of the fact that such general norms can be grasped only through the analysis of the capitalist economy, and not by induction of what seems to us to be common to all societies from mere empirical observation. See also the explanation by Sekine (Uno, 1980, 171).
2. This point has been confirmed repeatedly in the course of the controversy. See Yamaguchi (1987, 144–5), Miwa (1988, 34–5), Kobayashi (1981, 17–23), Itoh (1988, 133–6), Obata (1984b, 49–51), Kamakura (1970, 294–311). See also Haruta (1974). In particular Obata points out that the condition of same money wage rate should be necessary. Mawatari (1970) considers Uno to have failed in his demonstration of "the exchange of commodities at value" in the sense of "the exchange of commodities embodying equal amount of labour time" even in the absence of surplus labour owing to the lack of that condition.
3. Those whom I regard as belonging to the Suzuki group are as follows: Suzuki (1960/62), Ouchi (1964), Furihata (1965), Iwata (1967), Sakurai (1968), Hidaka(1983).
4. See Takasuka (1979, 114–19), Nagatani (1981) for a detailed characterisation of Suzuki group and comments against it.
5. See also Obata (1984a,b).
6. Nagatani (1981, 12) insists that there exists a competition of capitals for the rate of surplus value in the doctrine of production.
7. Okishio (1977, 26) thinks of a competition for the rate of income, which means the income divided by living labour, in a society of simple commodity producers.
8. These two articles appear as Chapters 2 and 3 in this book.
9. One feature of Sekine's labour theory of value can be found in his emphasis on "socially necessary labour", which means not only technically necessary minimum labour per each product, but also quantitatively necessary minimum labour for the whole of the product in order to meet social demand. See also Watanabe (1975a, 57–8).
10. I happen to assume here "the balanced equilibrium growth path which is called the 'golden age' path by modern economists" (Morishima, 1973, 68). It should be noted that Morishima tries to define the rate of profit (in terms of money) in terms of value, using "the golden age or von Neumann output vector" (1973, 70). See also Takasuka (1979, 131–7). Marx suggests that the essence of capitalism exists in maximal expanded reproduction. He says as follows: "Accumulate, accumulate! That is Moses and prophets! . . . save, save, i.e., reconvert the greatest possible portion of surplus-value, or surplus-product into capital!" (Marx, 1971a, 558). "So far as his [the capitalist's] actions are a mere function of capital . . . his own private consumption is a robbery perpetrated on accumulation" (1971a, 555).
11. If fixed capital and turnover-time are now abstracted from, invested capital is the same as cost-price.

5 Unoist Approach to the Theory of Economic Crises

Tomiichi Hoshino

5.1 INTRODUCTION[1]

To concentrate on the study of economic crises and business cycles is among the favorite choices of Marxian economists. There has been a bulk of literature on these particular subjects since Marx's *Capital* was published. What is rather unbelievable, however, is that Marxian economics had not been able to clarify the fundamental cause of economic crises and the mechanism of business cycles satisfactorily before Kozo Uno's monumental achievement. Uno's crisis theory, at its most abstract level, addresses two basic questions. One is to show the fundamental cause of crises in the actual process of capital accumulation, and this occurs in the context of *the theory of profit*. The other is to show the concrete mechanisms that explain the outbreak of crises, and this occurs in the context of *the theory of credit*. Uno sought to prove, not the breakdown of capitalism, but the logical necessity of crises and business cycles in a purely capitalist economy. His crisis theory also gives us an important insight into the historical treatment of crises and business cycles,[2] even though an abstract theory must not be directly applied to a concrete-historical analysis, except by the mediation of a mid-range theory, i.e., Uno's so-called stage theory.[3]

The purpose of this paper is to study the fundamental cause of crises by way of examining the significance and limitations of Marx's and Uno's theories of economic crises. In *Section II*, as a preliminary to this study, I will present an overview of the formation of Marx's crisis theory, focusing on *Capital*. In *Section III*, I will criticize Marx's "excess commodity theory" by pointing out, following Uno, that it neglects the working of market mechanisms. I will also show that, while underconsumption and disequilibrium among industrial sectors are not the fundamental causes of crises, disequilibrium among industrial sectors can cause speculation, and that the rise of prices which it entails constitutes an essential part of the pure theory of crisis. In *Section IV*, I will examine Marx's "excess capital theory" and argue that the fundamental cause of crises lies in a decline in the general rate of profit due to the drying-up of relative surplus population. I will also point out that Marx's so-called "lying idle" of capital is another factor which lowers the general rate of profit at the peak of prosperity. In *Section V* I will show that Marx underestimates the constraint of the existing fixed capital on a rise in the organic composition of capital, which prevents us from recognizing the fundamental cause of crises. I will also supplement Marx's view with Uno's that, in order to take the constraint of the existing fixed capital into consideration, the theory of capital accumulation should be logically developed *after* the theory of circulation. In *Section VI* I will show that, in trying to prove the decline in the general rate of profit, neither Marx nor Uno pays sufficient attention to fluctuations of commodity prices. I will then advance the following thesis. In principle commodity prices rise at the peak of prosperity, owing to an increase in demand for consumption goods and the bottleneck in production which the shortage of labour-power causes. This rise in prices may prevent the general rate of profit from declining. However, not only the rise in money wage-rate but also other symptoms of the shortage of labour-power (or of full employment) such as the "lying idle" of capital and the impotence of "an economy of scale" would not permit the general rate of profit to go on rising. Finally in *Section VII*, I will bring this paper to a conclusion.

5.2 THE FORMATION OF MARX'S CRISIS THEORY

In the early days of his economic research, Marx believed that an economic crisis was a critical moment for a socialist revolution (H. Oh'uchi, 1966, Ch. 1). As is well known, the last section of his plan for the system of the "Critique of Political Economy" was entitled "World Market and Crises". For him, to demonstrate the necessity of economic crises in a capitalist society meant to show the necessity of socialist revolution. This boils down to the so-called breakdown theory. Whenever a crisis broke out, Marx looked forward to the outbreak of a revolution, but only in vain. After a lengthy period of research, however, Marx seems to have changed his view on this matter.

In 1867 he published Volume I of *Capital*, but he could not finish the other volumes in his lifetime.[4] After his death, an enormous pile of his manuscripts in preparation for *Capital* was left, which Engels edited into Volume II (published in 1885) and Volume III (published in 1894). The whole system of *Capital*, it turned out, was no longer an embodiment of "Capital in General" as in his plan, but rather a system of *Capital* in "the ideal average" (Marx, 1971b, 170; see also Mawatari, 1973a, 19). Some important subjects such as competition among capitals, the credit system, ground rent, and crises or business cycles, which were once excluded from the system of "Capital in General" were introduced into the system of *Capital*.

It was Marx's ultimate goal in *Capital* (1971a, 10) to clarify "the economic law of motion of modern society". For Marx "this law" did not merely refer to the historical fact that capitalism came into being, developed, and would decline. It also referred to the mechanism of capital that gave rise to periodic crises and business cycles. Of course, we cannot deny the fact that even in *Capital* Marx (1971b, 353) often stated reservations such as "the cycles in which modern industry moves—state of inactivity, mounting revival, prosperity, over-production, crisis, stagnation, state of inactivity, etc., which fall beyond the scope of our analysis". Nonetheless, he repeatedly referred to crises and business cycles in the three volumes of *Capital*. Marx (1971a, Chap. 3 & Chap. 25; 1967, Part II & III; 1971b, Chap. 15 & Chap. 30–2) elucidated many important aspects of business cycles and crises, such as the abstract possibilities of crises, two different modes of capital accumulation and their relations with the demand for labour-power, the relationship between the existence of fixed capital and periodic crises, fundamental causes of crises, crises and the function of the credit system, etc.

These two seemingly contradictory facts, according to Aozai (1978, 109–14, 117), can be interpreted rationally only as follows: the pure theory of crisis was introduced into *Capital*, whereas historically concrete analyses of crises were left outside its system. Thus, as H. Oh'uchi (1966, 37; see also Itoh 1980, Ch. 4) points out, Marx's crisis theory was anchored in the framework of pure capitalism. It was part of the analysis of its inner structure and autonomous law of motion of capital, and hence it drew a clear line of demarcation from the so-called breakdown theory of capitalism.

It must be pointed out, however, that, while *Capital* is no longer the system of "Capital in General" but rather the system of "the ideal average" which includes the fundamental theory of crisis, Marx's crisis theory has not a few theoretical defects. Undoubtedly the theory was strongly influenced by the framework of "Capital in General". For example, as Uno (1953, 189) points out, Marx failed to formulate systematic theories of crises and business cycles systematically in *Capital*, certainly not in the manner that enables everyone to see what they are. This fact must stem, as many Unoists (Uno, 1953; Ito, 1973; Mawatari, 1973a; Kurita, 1992) have asserted, from the insufficient development of the theories of competition and credit in *Capital*. It is, therefore, my task to reconstruct, following Uno, a pure theory of business cycles, by critically appropriating the basic theory of crisis as we find it in *Capital*.[5]

5.3 THE EXCESS COMMODITY THEORY

Let us now investigate the main points of Marx's crisis theory. The most important statement on the causes of crises is found in Chapter 15 of Volume III of *Capital*. This chapter, however, is among the most complicated and controversial ones. Marx enumerates four different factors as causing crises, yet they seem to be either unrelated or even contradictory to each other.

As is well known, Marx's crisis theory can be divided into two types, that is, into the "excess commodity theory" and the "excess capital theory". Let us examine, in the first place, the excess commodity theory.

Marx looked upon underconsumption by workers and disequilibrium among industrial sectors as plausible causes of crises. He explained these two factors in the famous passage as follows:

> The conditions of direct exploitation, and those of realizing it, are not identical. They diverge not only in place and time, but also logically. The first are only limited by the productive power of society, the latter by the proportional relation of the various branches of production and the consumer power of society. But this last-named is not determined either by the absolute productive power, or by the absolute consumer power, but by the consumer power based on antagonistic conditions of distribution, which reduce the consumption of the bulk of society to a minimum varying within more or less narrow limits. It is furthermore restricted by the tendency to accumulate, the drive to expand capital and produce surplus-value on an extended scale. This is law for capitalist

> production, imposed by incessant revolutions in the methods of production themselves, by the depreciation of existing capital always bound up with them, by the general competitive struggle and the need to improve production and expand its scale merely as a means of self-preservation and under penalty of ruin. The market must, therefore, be continually extended, so that its interrelations and the conditions regulating them assume more and more the form of a natural law working independently of the producer, and become ever more uncontrollable. This intercontradiction seeks to resolve itself through expansion of the outlying field of production. But the more productiveness develops, the more it finds itself at variance with the narrow basis on which the conditions of consumption rest (Marx, 1971b, 239–40).

The same view is given in Chapter 30 of Volume III of *Capital.* Supposing that price fluctuations, the credit system, and speculation are all disregarded, Marx asserted:

> Then, a crisis could only be explained as the result of a disproportion of production in various branches of the economy, and as a result of a disproportion between the consumption of the capitalists and their accumulation. But as matters stand, the replacement of the capital invested in production depends largely upon the consuming power of the non-producing classes; while the consuming power of the workers is limited partly by the laws of wages, partly by the fact that they are used only as long as they can be profitably employed by the capitalist class. The ultimate reason for all real crises always remains the poverty and restricted consumption of the masses as opposed to the drive of capitalist production to develop the productive forces as though only the absolute consuming power of society constituted their limit (Marx, 1971b, 472–3).

Marx therefore stresses these two factors, the underconsumption of working-class and disequilibrium among industrial sectors, as being important causes of crises. However, the validity of these two cases of the excess commodity theory must be questioned.

5.3.1 The Underconsumption Type of Crisis Theory

Let us first examine the underconsumption theory, which is one variant of the excess commodity theory. Karl Kautsky, Rosa Luxemburg and

Nikolai Bukharin were all deeply influenced by this type of crisis theory (see Itoh, 1980, Ch. 5). It is certain that underconsumption by the working-class constitutes one of the important characteristics in capitalism. Marx, however, did not clearly explain why underconsumption necessarily entails over-production. Even if it does, the problem can actually be solved without resulting in a crisis. Now assume that an over-production of commodities arises in Department II, which produces consumption goods, because of underconsumption by the working class, so that capital and labour invested in this department is shifted to Department I, which produces means of production. The fluctuation of prices and differences in the rates of profit among various industrial sectors would bring about such a migration of capital and labour. As a result, the aggregate social labour would be re-allocated among these branches. In other words, the underconsumption of the working class can be overcome by the built-in market mechanism of a capitalist economy. Thus, underconsumption as such does not deter capital accumulation. Moreover, even if underconsumption causes a crisis, it fails to explain *periodic* crises. It is rather an important characteristic of capitalism that crises arise just when the consumption of workers expands due to a rise in real wages.

In fact, Marx himself later criticized the underconsumption theory and pointed out its fallacy as follows:

> It is sheer tautology to say that crises are caused by the scarcity of effective consumption, or of effective consumers. The capitalist system does not know any other modes of consumption than effective ones, except that of *sub forma pauperis* or of the swindler. That commodities are unsaleable means only that no effective purchasers have been found for them, i.e., consumers (since commodities are bought in the final analysis for productive or individual consumption). But if one were to attempt to give this tautology the semblance of a profounder justification by saying that the working-class receives too small a portion of its own product and the evil would be remedied as soon as it receives a larger share of it and its wages increase in consequence, one could only remark that crises are always prepared by precisely a period in which wages rise generally and the working-class actually gets a larger share of that part of the annual product which is intended for consumption. From the point of view of these advocates of sound and 'simple' (!) common sense, such a period should rather remove the crisis. It appears, then, that capitalist production comprises conditions which

> permit the working-class to enjoy that relative prosperity only momentarily, and at that always only as the harbinger of a coming crisis (Marx, 1967, 414–15).

This statement is found in Part III, Volume II of *Capital* where the theory of social reproduction is treated. Having formulated this theory, Marx may have clearly understood that the underconsumption of workers does not necessarily prevent capital accumulation from advancing because the aggregate demand for commodities are composed of both the demand for consumption goods and the demand for means of production. When the demand for consumption goods is reduced, some portion of the capital and labour invested for the production of these goods would be shifted out of this department to the other department which supplies the means of production.

5.3.2 The Disequilibrium Type of Crisis Theory

It cannot be denied that Marx criticized such an underconsumption theory, and he tended to place some stress on the disproportion theory among industrial sectors. For instance, in Chapter 20 of Volume II he referred to the "maintenance of fixed capital". According to him, one of the conditions for reproduction in a capitalist society to progress smoothly is that the replacement of the wear and tear portion of fixed capital in the form of money, and the replacement of fixed capital in kind, keep the balance in value terms with each other; if this condition is not kept, a crisis would arise: "There would be a crisis – a crisis of overproduction – in spite of reproduction on an unchanging scale" (Marx, 1967, 472; see also 472–3). Furthermore, in the next chapter he explained the conditions of social reproduction by saying that "inasmuch as only one-sided exchanges are made . . . the balance can be maintained only on the assumption that in amount the value of the one-sided purchases and that of the one-sided sales tally". He further asserted that these conditions would be changed "into so many possibilities of crises". Why are these conditions not secured? His answer to this question was "since a balance is itself an accident owing to the spontaneous nature of this production" (Marx, 1967, 498–9). In other words, since a capitalist mode of production is by nature anarchic.

Many Marxian economists have based their crisis theories on this view by Marx. Tugan-Baranovsky was one of the representatives of the disproportion theory. Tugan-Baranovsky (1972, Part II, Ch. 1) asserted that there is no other obstacle to the capital accumulation than disproportional allocation of social production caused by the anarchic nature of capitalism. Incidentally, he denies almost completely that a scarcity of labour-power would be a menace to entrepreneurs, since an industrial reserve army does not disappear even at the peak of prosperity.

However, the fact that capitalism is anarchic does not necessarily imply that it has no law. The greatest weakness of disproportionalists is that they underestimate the market mechanism that would adjust prices and redress the balance between demand and supply. Certainly, it might be possible that a crisis would occur because of a serious delay in adjusting the imbalance. But, such a disturbance does not necessarily arise periodically nor generally in the main branches of production of a society, except in the following two cases. (1) The delay in the correction of a disequilibrium or imbalance can appear when relative surplus population tends to dry up, especially towards the end of the phase of prosperity. I will discuss this case in the following subsection C. (2) The existence of huge advances of fixed capital in heavy industries would, in all likelihood, prevent the swift correction of a disequilibrium from occurring. But, such cases should be treated, following Uno's lead, at a more concrete level of analysis, that is in the stage theory of capitalist development or in the concrete-historical analysis of actual capitalism, which lies beyond the scope of this paper.

5.3.3 Uno's Criticism of the Excess Commodity Theory

As I have just shown, neither underconsumptionists nor disproportionalists have succeeded in demonstrating the periodicity of crises. Undoubtedly Uno was the most forceful critic of the excess commodity theory, and stressed the positive role of the market mechanism. For example, he states in one of his main works, *Kyoko-Ron* (Economic Crisis Theory) as follows:

> In a commodity-economy, such as the capitalist one, the capitalists ensure society's reproduction-process while making private decisions in light of prevailing prices over what kinds of commodities they should produce, i.e., commodities of particular use-values that are in social demand, and by exchanging them for one another. Exchange processes are, therefore, not likely to proceed smoothly. Some commodities are oversupplied while others are undersupplied at any

given set of prices. However, such imbalances are corrected by the fluctuation of prices. The equation of demand and supply is, of course, accomplished in the following manner: the output of the commodities whose prices decline will shrink and the output of the commodities whose prices rise will grow. Because capital is invested in various branches of production for the purpose of earning a maximal profit, this mechanism is established as commodity-economic necessity. Even though the aggregate reproduction process of capitalism is complex indeed, we must understand that while factors which intensify the imbalances of demands and supplies increase on the one hand, there also comes into being on the other hand the equilibrating mechanism which corrects such imbalances. Accordingly, if, in explaining the possibility of crises, we one-sidedly over-emphasize the fact that the capitalist mode of production is blind and anarchic, we will lose sight of the other side of the same mode of production, and fail to recognize its true nature. Capitalism is more resilient than is sometimes made out to be (Uno, 1953, 60–1).

Since the "capitalist mode of production is an anarchic and blind production", there will arise an incessant disequilibrium among various branches. That is true. However, with the working of the market mechanism, there is no reason for us to expect that this disequilibrium necessarily leads to a sharp disturbance in the reproduction-process of the aggregate social capital (Sekine, 1986, 270). Migration of capital and labour among various sectors of production, accompanied by movements of commodity prices, will support the working of this mechanism. Were it not for this crucial mechanism, the reproduction-process of capital would be thrown into chronic depression and the capitalist mode of production would soon cease to exist.

In addition, as Uno (1980; see also Sekine 1984) stresses, the theory of reproduction-schemes is not meant to be used for the study of crises. According to him, the purpose of the theory of reproduction-schemes is to show how, as in all societies, the means of production and consumption goods are reproduced under capitalism as well, and how the supra-historic conditions of reproduction are maintained provided that the adequacy of the supply of labour-power is guaranteed. Nevertheless, it has been traditionally the popular practice among Marxian economists to apply the theory of reproduction-schemes, following Marx, to the crisis theory. This was based on their failure to understand the meaning of the theory of reproduction-schemes. Tugan-Baranovsky (1972) and Grossmann's views (1932) were the most typical examples of "the wrong interpretations and the misuses of the reproduction-schemes" (Uno, 1980, 69; see also Sekine, 1984, 484). In relation to this point, Albritton (1986, 58) calls our attention to the fact that "the reproduction schema occur at the end of Volume II and not the end of Volume III of *Capital.*" As he contends, the reason for this is "because Marx never thought such schema could possibly serve to encapsulate the capitalist mode of production as a whole".

As Uno contends, the subject-matter of the theory of reproduction-schemes is to show how capitalist society accomplishes, by the working of the law of value, the same social conditions for reproduction which other societies in history accomplish as well. Uno explains this point as follows:

> In all societies the continuity of annual reproduction depends on the allocation of both labour-power and the means of production to the two sectors of production in accordance with the degree of intensity with which the products of each department are socially required. A capitalist commodity-economy satisfies this norm by the law of value which asserts itself through the movement of prices (Uno, 1980, 55).

Although the law of value is specific to capitalist society, what the law accomplishes is suprahistoric. Accordingly Uno calls the social conditions for reproduction "the absolute foundation of the law of value" (Uno, 1980, 54). Thus, it is certain that, to him, the theory of reproduction-schemes is not meant to show the fundamental cause of crises. As will be shown later, this fallacy comes in part from the imperfect treatment of the theory of reproduction in *Capital.*

Granted that the above is correct. Should we then refuse to take into consideration any kind of disequilibrium as not relevant to the outbreak of crises? Uno's answer is positive.[6] It is on this point that I beg to differ.

In the presence of advances of fixed capital in the production process, it is not possible to expect achievement of demand and supply balances in the short run by the direct migration of capital and labour from one industry to another. We must look for a more concrete process of the working of the market mechanism. Under simple reproduction, imbalances cannot be corrected until such time as the existing fixed capitals must be replaced. Under expanded reproduction, however, the corrective mechanism works more easily, provided that the supply of labour-power is abundant. For it can be accomplished without direct migration of capital and labour from one industrial sector to another. Since, in the oversupplied industries, the rates of profit are lower than the general rate, they will grow much more slowly than the rest of the economy. Likewise, in the undersupplied industries, the rates of profit must be higher than the general rate, so that they are bound to grow much faster than the rest of the economy.

However, this mechanism of establishing and maintaining balances among different branches of production works only when there exists an abundant supply of labour-power. If a relative surplus-population dries up because of a rapid progress of capital accumulation, which will necessarily be the case at the peak of prosperity, this mechanism will be stymied. Disequilibrium among industrial sectors will persist, at least for a short period of time at the peak of prosperity. That disequilibrium touches off a fever of general speculation at this point, as I will show later. Thus, even though the disequilibrium itself is not a fundamental cause of crises, it can count as one significant element in the framework of the excess capital theory.[7]

5.4 THE EXCESS CAPITAL THEORY

Now let us turn to Marx's excess capital theory. Marx regarded a decline in the general rate of profit as one of the most important causes of crises. But, how does the rate of profit decline? Marx enumerated two types of decline in the general rate of profit: one is a decline caused by a rising organic composition of capital, and the other is the one caused by a sharp rise in wages owing to a shortage of labour-power. These two types of decline must be separated one from the other, since they are logically unrelated to each other. Of course, Marx knew it well. Nevertheless, he treated them together in the same chapter, and by doing so his treatment has become quite misleading.

5.4.1 The Law of the Falling Rate of Profit and Crises

Let us begin by taking note of the title of Chapter 15 in Volume III of *Capital*, which is "Exposition of the Internal Contradictions of the Law". This gives us the impression that in this chapter he intends to describe crises as the expression of the internal contradictions of the law of a tendency for the rate of profit to fall. Moreover, Marx (1971b, 237) points out, in a paragraph of Section 1 of Chapter 15 in Volume III, that "a fall in the rate of profit" caused by the accumulation which is accompanied by a rising organic composition of capital would breed "over-production, speculation, crises, and surplus-capital alongside surplus-population".

Not only the orthodox school but also Uno (1980, 87), Okishio (1976, Ch. 4), and Sekine (1986, Part III, Ch. 7) support this law of Marx.[8] However, I question whether such a law does indeed hold good, for the following reason. Even if the organic composition of capital rises as a result of advancement in productivity, both a rising rate of surplus-value and an increasing frequency of turnover of capital may counterbalance the tendency for the rate of profit to fall. It is from this point of view that Sweezy (1967, Ch. 6) denies this law. Also many Unoists in Japan reject this law.[9] Moreover, I wonder *whether it is proper to assume that a rising technical composition of capital would be necessarily accompanied by a rising value-composition of capital.* An advancement in productivity may lower the value of means of production. Hence, even if a technical ratio of means of production to labour-power rises, the ratio of constant capital to variable capital does not necessarily increase. In other words, a rising technical composition of capital is not necessarily accompanied by a rising organic composition of capital.[10]

In addition, setting aside the question about whether the so-called law holds good or not, the tendency for the general rate of profit to decline is nothing more than a historical tendency which is observed over a period of time exceeding the length of a business cycle. Hence it cannot be a fundamental cause of crises. Therefore, the title of the Chapter 15 should not be "Exposition of the Internal Contradictions of the Law", but rather be "Exposition of the Internal Contradictions of the Capitalist Mode of Production", following Uno and Sekine (Uno, 1953, 220–1 and 1980, 94–5; Sekine, 1986, Part III, Chap. 7).

5.4.2 The Absolute Over-Production of Capital

(a) Marx's theory of the absolute excess of capital is the other type of his crisis theory. In Section 3 "Excess Capital and Excess Population" of the same chapter, Marx emphasizes that advancement in productivity will bring about excess capital and excess population. Marx talks of an over-production of capital, and says that "over-production of capital always includes over-production of commodities" (Marx, 1971b, 246). It is noteworthy here that the over-production of capital is regarded not as a result, but rather as the cause of the over-production of commodities. Therefore, the "over-production of capital" is a key concept for understanding the theoretical necessity of crises. In order to clarify this point, Marx (1971b, 245–6) starts his argument assuming the "absolute over-production of capital". This means that value-augmentation becomes impossible not just "in one or another, or a few important spheres of production," but absolutely "in its full scope", i.e., in "all fields of production".

More concretely, the absolute over-production of capital implies that "the increased capital $C + \Delta C$ would produce no more, or even less, profit than capital C before its expansion by ΔC" (Marx, 1971b, 246). For example, let us assume that while the total capital of 100 units once produced 10 units of profit, now an increased amount of total capital of 120 units produces 10 or less units of profit. So long as the marginal portion of capital gains positive profit ($\Delta S > 0$), the credit system not being taken into consideration, capital accumulation can proceed further (Sekine, 1984, 551). In this case, however, the marginal capital of 20 units produces zero or negative profit. If marginal profit is zero or negative ($\Delta S \leq 0$), capital accumulation becomes meaningless for capitalists, and the general rate of profit falls suddenly in reflection of this fact. I must now explain the exact manner in which such a fall occurs.

It is noteworthy that Marx has clearly identified the true reason for the general rate of profit to fall. It is not because of a change in the organic composition of capital "caused by the development of the productive forces", but because of a sudden rise in money wage-rate caused by the over-accumulation of capital. As this latter exhausts the supply of labour-power, the general rate of profit falls suddenly (1971b, 246–7). The following may be remarked in passing. It is not realistic to assume that a rise in wages would at once entail a substitution of machinery for labour-power in a short period, for example, at the peak of prosperity.[11]

What is characteristic of labour-power as a commodity is that the demand for labour-power does not necessarily lead to its increased supply because it is not a product of capital, but is only reproduced in the family life of the working class. Therefore, the shortage of labour-power would cause a particularly serious obstacle to capital accumulation. The excess of capital reflects "a more fundamental disequilibrium" (Sekine, 1984, 555) than imbalances among ordinary commodities; as Uno (1953, 123) also stresses, the former disequilibrium cannot be overcome by the market mechanism. There arises, therefore, a general over-production of commodities, which does not originate merely in the commodity-form. Therefore, this type of "excess capital theory" by Marx is a key to the demonstration of the logical necessity of crises in a pure capitalism.

Now let us follow the logic of his theory in Section 3 of Chapter 15, Volume III, step by step. Here Marx focuses on the very specific phenomenon of the "lying-idle of capital" which attends an absolute over-production of capital.

(b) What would happen if an absolute over-production of capital occurs? Marx (1971b, 247) himself replies as follows:

> In reality, it would appear that a portion of the capital would lie completely or partially idle (because it would have to crowd out some of the active capital before it could expand its own value), and the other portion would produce values at a lower rate of profit, owing to the pressure of unemployed or but partly employed capital.

However, what does such a state of things as "lying idle of capital" mean concretely? It seems quite difficult to know what Marx had in his mind when he referred to the "lying idle" of capital. Almost no Marxian economist, not even Uno, seems to have successfully clarified what Marx meant by that term. Moreover, most Unoists are, following Uno, quite critical of this explanation by Marx. Uno's main point of criticism of Marx is that if there exists severe competition among capitals, even a portion of capital cannot "lie idle" until crises break out (1953, 223–5). Although I once agreed with Uno on this point (Hoshino, 1977, 33–4), I now hold a different view. I consider that once a pool of a relative surplus-population dries up, some portion of capitals will be forced to "lie idle" even before a crisis breaks out. My interpretation of his view is as follows.[12]

The starting point of Marx's argument is that there exists full employment. In other words, Marx (1971b, 246) is assuming the utmost limits where "capital would . . . have grown in such a ratio to the labouring population that neither the absolute working-time supplied by this population, nor the relative surplus working-time, could be expanded any further (this last would not be feasible at any rate in the case when the demand for labour were so strong that there were a tendency for wages to rise)". If the accumulation of capital continues even after full employment is achieved, capitalists will scramble for labour-power and will even try to entice workers from other sectors or firms. As a result, the money wage rate will have to rise rapidly. In inverse proportion, however, the general rate of profit will decline. This is how Uno stated the case.

However, what would happen to the capitalists who were defeated in their effort to entice workers from others? *Some portion of their capitals would be forced to lie idle even if capitalists have a strong desire to accumulate.* Some old or newly installed fixed capitals could no longer operate because of a scarcity of labour-power. In other words, the socially aggregate output could no longer expand on the condition that in the phase of prosperity capital accumulation advances in the presence of a constant organic composition of capital. Moreover, it is the competition or struggle among capitalists that will decide which portion of capital would lie idle,[13] as Marx says:

> A portion of the old capital has to lie unused under all circumstances; it has to give up its characteristic quality as capital, so far as acting as such and producing value is concerned. *The competitive struggle would decide what part of it would be particularly affected.* So long as things go well, competition effects an operating fraternity of the capitalist class, as we have seen in the case of the equalization of the general rate of profit, so that each shares in the common loot in proportion to the size of his respective investment. But as soon as it no longer is a question of sharing profits, but of sharing losses, everyone tries to reduce his own share to a minimum and to shove it off upon another. The class, as such, must inevitably lose. *How much the individual capitalist must bear of the loss, i.e., to what extent he must share in it at all, is decided by strength and cunning, and competition then becomes a fight among hostile brothers.* The antagonism between each individual capitalist's interests and those of the capitalist class as a whole, then comes to the surface, just as previously the identity of these interests operated in practice through competition (1971b, 248. The emphasis is mine).

Thus, a portion of capital must inevitably "lie idle." Hence the general rate of profit will decline still more owing to such "lying idle of capital" along with a rise in money wage rate. This is the case which Marx (1971b, 247) described as follows: "The rate of profit would not fall under the effect of competition due to over-production of capital. It would rather be the reverse; it would be the competitive struggle which would begin because the fallen rate of profit and over-production of capital originate from the same conditions." Thus, the absolute over-production of capital will go beyond the limits and will be the fundamental cause of crises.

(c) Marx also elucidates how the economy recovers from its stagnation. He enumerates the factors that would reverse the declining rates of profit as follows (Marx, 1971b, 248–50): (i) the "destruction" and "depreciation" of capitals in their value or substance at the phases of crises and depressions; (ii) the lay-off of a part of the working-class and "a reduction of wages below the average"; (iii) the lowering of "the individual value of the total product below its general value by means of new machines, new and improved working methods, new combinations, i.e., to increase the productivity of a given quantity of labour, to lower the proportion of variable to constant capital, and thereby to release some labourers"; (iv) "the depreciation of the elements of constant capital". Marx points out that the methods of production would be inevitably innovated at the phase of depression under the pressure of "the fall in prices and the competitive struggle" (Marx, 1971b, 250). And their motivation to introduce new methods of production is to have an extra surplus-value, as Marx proves in Volume I.

Uno (1953, 79–81, 135–46) must have highly appraised Marx's view of this matter. According to him, the excess of capital implies that a value-relation between capital and labour has become untenable for value-augmentation, and the phase of depression lasts until the capital-labour relation is renovated. Moreover, the above factor (iii) is a decisive one for its renovation (1953, Chap. 2 & 3).

"And thus the cycle would run its course anew" (Marx 1971b, 250). Judging from our investigation up to this point into Marx's excess capital theory, it can be properly concluded that whereas Marx has not completed his crisis theory in details, he has given a broad outline of the business cycle in this section. The theory of crises and business cycles at the level of pure theory should be established, following Uno, on the basis of "the absolute over-production of capital".

Nevertheless, it is still not so clear whether Marx in fact recognized that "the absolute over-production of capital" is logically necessary and periodically appears in the course of business cycles, since Marx seems to assume a constant rise in the organic composition of capital.

Uno (1953, 222–3; see also Itoh, 1988, 295) criticized Marx by claiming that Marx (1971b, 250) argues the absolute over-production of capital "under extreme conditions". If so, the important obstacle of the shortage of labour-power for capital accumulation will be incessantly remedied and capital will accumulate indefinitely. Then, the fundamental cause of crises will be lost sight of. This defect goes back to the logical difficulties in his theory of capital accumulation in Volume I.

5.5 THE CAPITALIST LAW OF POPULATION AND CRISES

(a) In the theory of capital accumulation, Marx (1971a, Ch. 25) explicated two different modes of capital accumulation: accumulation accompanied by a constant organic composition of capital and accumulation accompanied by a rising organic composition of capital.[14] In addition, he formulated the capitalist law of population which is quite distinct from Malthus' natural law of population. However, on the one hand, Marx overestimated the possibilities for accumulation under a rising organic composition, or a "deepening" accumulation in contrast with accumulation under constant organic composition of capital, or a "widening" accumulation. He thought that capital accumulation usually advances with a constant rise in an organic composition of capital: capital is accumulated "under a progressive qualitative change in its composition, under a constant increase of its constant, at the expense of its variable constituent" (1971a, 628–9). On the other hand, he views the capitalist law of population one-sidedly from the standpoint of a progressive formation of a relative surplus-population, thinking lightly of its absorption.[15] This eventually leads to Marx's so-called thesis of the absolute or relative impoverishment of the working-class. Of course, the latter problem is the other side of the coin of the former. Let us consider the reasons for these problems.

These problems of his theory originate, as Uno (1962, 311–13) says, from the fact that Marx underestimated an important constraint caused by existing fixed capitals which prevent a constant innovation in the methods of production and an accompanying rise in the organic composition of capital.

Marx (1971a, 627–8) gives two reasons why the organic composition of capital rises constantly. One of the reasons is that the centralization of capital "extends and speeds those revolutions in the technical composition of capital which raise its constant portion at the expense of its variable portion"; and, he claims that joint-stock companies are very useful for it. The other reason is as follows:

> The additional capitals formed in the normal course of accumulation . . . serve particularly as vehicles for the exploitation of new inventions and discoveries, and industrial improvements in general. But in time the old capital also reaches the moment of renewal from top to toe, when it sheds its skin and is reborn like the others in a perfected technical form (Marx 1971a, 628).

However, even if the centralization of capital and the introduction of new inventions by additional capitals speed up "deepening" accumulation, once constructed, fixed capital can not be instantly replaced by new fixed capital. Its replacement will be impossible until the wear and tear of fixed capital is accomplished. Setting aside the portion of additional capital, "deepening" accumulation cannot advance progressively without sacrificing the existing fixed capital. Therefore, a "widening" accumulation is still a fundamental aspect of accumulation. As a result, the demand for labour-power will also expand along with a "widening" accumulation. Of course, as Marx remarks, additional capitals may serve to introduce "new inventions and discoveries." However, it should be noted that these additional capitals produce a new demand for labour-power, although they "attract fewer and fewer labourers in proportion to its magnitude" (Marx, 1971a, 628).

Uno examined the reason why Marx neglected the constraint of the existing fixed capital on the otherwise constant rise in the organic composition of capital and found it in the logical composition of *Capital*: the theory of "The Accumulation of Capital" is developed in Volume I *before* the theory of "The Process of Circulation of Capital" in Volume II, which includes the chapters on "Fixed Capital and Circulating Capital" (1971a, 313–14). Instead of this sequence of categories, Uno proposes a new composition of the system of the principles of political economy, where the whole system is divided into three parts, i.e., Part I on "The Doctrine of Circulation", Part II on "The Doctrine of Production" and Part III on "The Doctrine of Distribution"; and Part II is also divided into three chapters, i.e., Chapter 1 on "The Production-Process of Capital", Chapter 2 on "The Circulation-Process of Capital" and Chapter 3 on "The Reproduction-Process of Capital"; Chapter 3 is composed of two theories of "The Accumulation of Capital" and of "The Reproduction-Process of the Aggregate Social Capital (so-called "The Theory of Reproduction Schemes"). Thus, Uno's *Principles* clearly shows the constraint of the existing fixed capital on the constant rise in the organic composition in the system itself.

Moreover, taking this constraint into consideration, Uno takes the relationship between Marx's so-called two modes of capital accumulation as taking turns in principle, i.e., a "widening" accumulation in a prosperity phase and a "deepening" accumulation in a depression phase. In a "widening" phase, the labour-power is absorbed by degrees into industrial sectors and a pool of relative surplus-population finally dries up; contrariwise, in a "deepening" phase labour-power is expelled from industrial sectors and the pool of surplus-population swells up. In addition, a rise in the organic composition will appear as a severe competition among capitalists to survive the depression. And this rise will supplement the pool of surplus-population. Thus, Uno depicts the capitalist law of population in its two aspects. And Marx's so-called "the absolute over-production of capital" appears periodically at the peak of a "widening" phase. Marxian economic crisis theory is now given a secure basis by Uno's *Principles*. Thus, Uno (1953; see also Sekine, 1984; Albritton, 1986, 54–8; Itoh, 1980 and 1988) gave a significant solution to this problem.

However, there still exists another flaw in Marx's statement mentioned above: when he tries to prove a sudden fall in the general rate of profit caused by a rise in money wage rate, he pays no attention to a rise in the prices of commodities. This is methodologically fallacious because the amount of profit is defined not by the money wage rate, but by the real wage rate, the latter of which is also defined by both the money wage rate and the prices of consumption goods; otherwise he assumes the prices of commodities are equal to their value-prices. Theoretically speaking, however, the prices of commodities rise at the phase of prosperity, as will be shown later. Unless a rise in prices of commodities is taken into account, it can not be said that the necessity of the absolute over-production of capital is demonstrated in the actual process of capital accumulation. Whereas Uno noticed this difficulty, he did not solve it. In the next section, I will examine Uno's view on this point and also try to give my own solution to it.

5.6 A RISE IN MONEY WAGE RATE VS. A RISE IN PRICES

(a) As Mawatari (1973, 36–7) argues, Marx looks at wages only from the cost-side and almost completely neglects the demand-side of the story. On the other hand, Uno (1953, 57–8, 91–7) actually refers to the speculative elevation of commodity prices during the phase of prosperity, and admits that a sharp rise in prices may well cover up the decline in the general rate of profit. Yet he insists that this does not reflect a true value-relation between capital and labour, but only an illusionary one. Thus, according to Uno (1953, 93), "the tendency of commodity prices to rise, except in the case of the rise in the price of labour-power, must remain within limits since the price rise by itself entails an increase in the supply, so that balances between demands and supplies will be restored". He further points out as follows:

> Actually the activities of commercial capital . . . add considerable distortion to the normal development of the reproduction-process during the phase of prosperity. For by speculatively pushing up certain prices, they conceal the true price relations of the reproduction system. The prices of certain commodities thus continue to rise even when their unsold stocks build up. For the theoretical clarification of the cause of capitalist crises, it is, therefore, best to set aside the activities of commercial capital, and that is the position we maintain in this book. However, for the detailed description of the cyclical phases of prosperity, crisis and depression, commercial capital along with merchant capital do play important roles to which I intend to pay attention, even though their roles distort the general price trend and thus make us lose sight of the more fundamental relations (1953, 57–8).

I, however, differ from Uno in claiming that neither the rise in prices nor speculative activities of commercial capital are "the factors that would make us lose sight of the fundamental relations", but are rather well-grounded in the reproduction process at the phase of prosperity. Therefore, even when we approach the pure theory of crises, I do not believe that "commercial capitals should be excluded from the theory [of crises]" (see Takei, 1985, 52–3).

(b) Let us consider why commodity prices rise in the phase of prosperity along with a rise in money wage-rate.

As capital accumulation advances under a constant organic composition of capital in the phase of prosperity, in the manner already shown, the demand for labour-power would increase, resulting in the exhaustion of a relative surplus-population. Consequently, the money wage-rate would rise sharply. However, the prices of commodities usually would also rise during this phase.

Generally speaking, it is the difference between the formation and the expenditure of funds that gives rise to a fluctuation of prices. Here, the formation of funds implies the sale or the supply of commodities and the expenditure of funds for the purchase of or demand for commodities. Commodity prices would rise slowly from the end of a depression to the phase of prosperity because the expenditure of funds exceeds the formation of funds (i) as a result of a general replacement-investment for an innovation in the method of production at the end of a depression and also (ii) as a result of a new investment in the phase of prosperity.[16] Therefore, the credit system itself is not the principal cause of the rise in prices during this phase, though it may merely accelerate the rise.[17]

In addition, commodity prices would begin to rise sharply around the peak of prosperity as the supply of relative surplus-population begins to be exhausted. For, the shortage of labour-power will raise the money wage-rate, and it will also give birth to a bottleneck in production. A rise in the money wage-rate indeed raises the production cost for capitals, but it also increases the demand of the working-class for consumption-goods because they can now afford to spend more money on them. We also need to take the supply-side effect of full employment into account; the aggregate-social output of commodities can no longer expand smoothly because of a scarcity of labour-power. In other words, the supply of commodities tends to lag behind the increase of demand. Thus, while a rise in the money wage-rate increases the demand for consumption-goods, output can no longer expand quite as rapidly as before, because of the shortage of labour-power. Moreover, it is not easy to extend the working-time, especially when the demand for labour-power is rising.

As Marx (1971b, 246) states, "neither the absolute working-time supplied by this population, nor the relative surplus working-time, could be expanded any further (this last would not be feasible at any rate in the case when the demand for labour were so strong that there were a tendency for wages to rise)". As a result, the prices of commodities would increase rapidly. In addition, the speculative stockpiling of these commodities by commercial capital would also be active because such a rise in prices would encourage the belief that the rate of profit would rise in the near future. Therefore, prices would increase more rapidly in this phase.

(c) If we take such a rise in prices and speculations into account, we must notice that there arises another difficulty at this point: how to show a sudden decline in the general rate of profit in the phase of prosperity.[18] However, whether or not the rate of profit would sharply decline because of the shortage of labour-power, it should be noted that the mechanism of raising the rate of profit during the phase of prosperity would not work any more for the following reasons.

First, fixed costs per unit commodity tend to increase, since, while the total fixed costs per capitalist increase owing to new investments in fixed capital during the prosperity phase, the production of commodities cannot expand so rapidly as before as a result of the exhaustion of the relative surplus-population. In other words, so-called "economies of scale" would be gradually lost. *Secondly*, the distribution ratio of the employed population between the two departments of social reproduction would no longer be higher for the first department because it becomes more and more difficult for the capitalists in the first department to expand employment as a result of the shortage of labour-power. It means that the real wage-rate would not go on declining any more. *Thirdly*, the entire time of turnover of a given capital would be extended at the peak of prosperity because, for one thing, the circulation-period becomes longer owing to the speculative stockpiling of commodities and, for another, the purchasing-period of labour-power as a commodity becomes longer as a result of the shortage of labour-power. *Lastly*, as Marx emphasizes, a portion of capitals will be forced to lie idle because of full employment. Some capitalists might not be able to start up a newly installed fixed capital owing to the difficulty of getting additional employment, or old fixed capitals may be forced to shut down in whole or in part.

These four factors imply that even if a sharp and sudden decline in the rate of profit could not be proved, at least the rate of profit would be relatively stagnant.

5.7 CONCLUSION

Marx's two types of crisis theory have given rise to considerable confusion and to many sterile controversies for a long time in the history of Marxian economics. Marxists and Marxian economists have been dazzled by Marx's "excess commodity theory". They were undoubtedly fascinated by the idea that the underconsumption of the labouring-class and the disequilibrium caused by the anarchistic character of capitalism are inner contradictions of capitalism. Moreover, this idea also conforms to their socialist ideology. The "excess capital theory", in contrast, seemed to them to be "non-socialist" or otherwise unsatisfactory. For example, it was inconvenient for them to organize labour movements on the basis of the theory which implies that a rise in wages causes crises, and that the main evil of capitalism can be remedied if they abstain from demanding a rising wage-rate. They seem to have forgotten that surplus-value is the deduction of the value of labour-power from newly formed value by labour.

Uno's crisis theory is certainly one of the most fundamental achievements in the history of Marxian economics because it puts an end to such misconceptions. Uno "purified" Marx's crisis theory on the basis of "the absolute over-production of capital" which we find in Volume III of *Capital*. On the one hand, he exposed the fallacy of the excess commodity theory, by emphasizing the significance of the price mechanism, and he gave a consistent and systematic explanation of crises and business cycles on the basis of the excess capital theory. He recognized the commodification of labour-power to be the main contradiction inherent in capitalism, and put Marx's so-called two modes of capital accumulation in sequential order, i.e., as taking turns in business cycles, and emphasized the working of "competition among capitals" for both intensifying the excess of capital at the peak of prosperity and forcing capitals to replace more or less depreciated fixed capitals with new ones at the phase of depression. Thus, Uno has overcome the opposition between Marx's two types of economic crisis theory.

However, I must admit that his theory has left a few theoretical difficulties unsolved. These difficulties lie for the most part in Marx's original crisis theory. One of the most important points is that Uno neglected the role of fluctuations in commodity prices in business cycles. Though he paid attention to the rise in prices at the peak of prosperity, he considered it to be of minor theoretical importance in contrast with the rise in wages, reasoning that a rise in commodity prices other than the rise in the price of labour-power would remain within limits, since the price mechanism works to restore the balance between demand and supply.

However, in showing the decline in the general profit rate, we must take into account the rise in the prices of commodities, since it is not the money wage-rate but the real wage-rate that determines the rate of profit, and commodity prices will necessarily rise owing to the increased demand for consumption-goods and speculative stockpiling of commodities on the one hand, and bottlenecks in production on the other. The rise in prices will more or less prevent the general rate of profit from declining in some sectors, the fact of which can mislead us into believing that the excess of capital does not as yet exist when in fact it is already present.

It cannot be denied, however, that capital accumulation in the phase of prosperity is limited by the full employment of labour-power. I also focus on a set of symptoms caused by the shortage of labour-power such as: (i) rising fixed costs per unit of output, (ii) reversal of the declining trend of real wage-rate, (iii) a longer period of time for the turnover of capital, and (iv) the "lying idle" of a portion of capitals. I call these symptoms: "full employment syndromes". These symptoms will necessarily work as factors that reduce profits when the money wage-rate rises. Therefore, the general rate of profit will be stagnant.

I have delayed mentioning an important problem until now; namely, the question of the periodic outbreaks of crises. Uno looked upon the credit system as the critical factor for the outbreak of crises. He asserts that, while the general rate of profit declines at the peak of prosperity, the rate of interest will rise sharply and that as the result of the conflict between them crises will break out. However, Uno did not succeed in showing how the creation of credit by the central bank breaks down at the peak of prosperity just when it is most needed. It is important for us to solve these problems in order to show the logical necessity of crises and business cycles. In my opinion, "full employment syndromes" will also tighten the money market and push up the rates of interest. This subject must, however, be postponed to another occasion.

Notes and References

1. I am deeply indebted to professors Tomohiko Sekine and Robert Albritton at York University for their kind supervision during my stay in Canada as a visiting scholar at York University from April 1991 to March 1992, as well as for many helpful comments on my first draft of this paper. I am also grateful for their editorial assistance given that the present is the first paper that I have written in English. I am, of course, responsible for all errors that may still remain. I wish to dedicate this paper to Professor Sekine in celebration of his 60th birthday.
2. Not a few Unoists have approached the analysis of stagflation from the viewpoint of Uno's crisis theory and achieved considerable success. I take that to exemplify the validity of Uno's crisis theory in the study of contemporary capitalism. See M. Itoh (1980), T. Oh'uchi (1983) and Mawatari (1977; 1979).
3. Although Uno first advanced the necessity of "stage theory", which is to mediate between pure economic theory and a concrete and historical analysis of capitalism, he did not fully develop it by himself. His book entitled *Keizaiseisaku-Ron* (The Types of Economic Policies under Capitalism) treats a fairly narrow range of stage-theoretic issues, focusing specifically on economic policies. However, the book provided the foundation of stage theory by explicating the typical form of capital dominant in each stage of capitalist development, as it operates in the leading industry of the leading country (Uno, 1971). The English translation of this book is in preparation. For a more concrete and fully systematic development of stage theory, see Albritton's ambitious book (1991). The Japanese translation of his book will be published in 1994.
4. Sweezy (1942, 134) asserts that "if Marx had lived to complete his analysis of competition and credit he would have given us a thorough and systematic treatment of crises". However, taking the many logical

difficulties present in *Capital* into account, I cannot be so optimistic. In any case, "if" has little importance in history.

5. Sekine (1984 and 1986) and Itoh (1980 and 1987) are the two main Unoists who have contributed, outside Japan, toward the development of crisis theory in the context of pure capitalism.
6. H. Oh'uchi (1977) decisively refuses to accept that a disequilibrium will play any role in the outbreak of a crisis. My position on this matter is that disequilibrium among industrial branches can coexist with the excess of capital. See also Note 7 of this paper.
7. H. Baba (1973 and 1972) is one of the pioneers of a logical extension of the excess capital theory who claim that a disequilibrium among industrial sectors constitutes an important factor in the outbreak of a crisis. See also Mawatari (1973b), Tohara (1972), Yamaguchi (1984), Sugiura (1975), Kobayashi (1979), Itoh (1973), Hoshino (1977) and Matsuda (1990). Once the logical necessity that a disequilibrium appears as a result of the shortage of labour-power is admitted, it (disequilibrium) cannot be ignored among the causes of crises. I do not, however, intend to assert that a crisis never breaks out unless a disequilibrium among industrial sectors first makes its appearance. In this regard, I am in agreement with Sekine (1986, 268), who says:

   > Indeed even in the absence of any significant inter-industry disproportionately or disequilibrium an excess of capital cannot be avoided when the widening phase of accumulation proceeds far enough. There can be no fundamental cause of periodic crises other than the decennial recurrence of the excess of capital.

8. While I do not agree with Sekine's conclusion that "the law of the falling rate of profit" can be demonstrated to be valid, I regret to say that his mathematical explanation of this law goes beyond my comprehension owing to my limited ability in mathematics.
9. M. Sakaguchi (1960) was perhaps the first in the Uno school to clearly criticize this law. Later Suzuki (1962), Hidaka (1983) and Yamaguchi (1985) concurred and excluded the law from their system of Principles, i.e., from their pure theory of capitalism. I agree with their treatment of this law because of my conviction that this law can neither be demonstrated nor can I see any significance of this law in the general theory.
10. Okuyama (1989, 174–5) also indicates the same viewpoint.
11. Sekine (1986, 254–5) sharply criticizes Marx (1965, 638) and Sweezy (1967, 108–9) for their assertion that a rise in wages would at once bring about a substitution of machinery for labour-power, by using a metaphor as follows:

    > But if Mr. Sweezy owns a large American car driven for only a few years and is suddenly faced with the rising cost of motor fuel, does he at once scrap his valuable vehicle and buy a new Toyota or Honda in order to punish oil companies, i.e., to "hold their pretensions in check"? Even if he personally has enough resources and fortitude to

> do so, should he expect other persons in the same condition to follow him *en masse* reducing the value of all large cars to practically nil and socially creating a relative surplus of motor fuel in no time? Such a thing cannot happen, nor does it make much economic sense.

And he concludes:

> For exactly the same reason capitalists who have invested their valuable fortune in fixed capital would not scrap it before it is sufficiently depreciated, even if they collectively (?) agreed to punish pretentious workers. That is why the excess of capital inevitably occurs as the "widening" phase of capital accumulation continues in the face of rising wages.

12. As far as I know, Kikuo Suzuki (1966, 124–34) and Yoshihiro Takasuka (1985, 200–1) more or less successfully interpreted what Marx meant. I here owe my interpretation to them.
13. Although I admit that Marx's exposition of an inevitable "lying idle" of a portion of capitals is reasonable, I also agree with Uno's criticism of Marx that it is not until a crisis breaks out that a widespread "lying idle" of capital arises. I furthermore agree with Uno's other criticism of Marx that the latter seems to explain "the competitive struggle" of sharing losses among capitals by invoking a behavior of monopolistic capitals as follows:

    > The part of ΔC in the hands of old functioning capitalists would be allowed to remain more or less idle to prevent a depreciation of their own original capital and not to narrow its place in the field of production. Or they would employ it, even at a momentary loss, to shift the need of keeping additional capital idle on newcomers and on their competitors in general (1962, 247).

    As Uno (1953, 223–5) critically pointed out, it is not reasonable to assume that competitive capitalists would let some portion of their capitals remain idle "to prevent a depreciation of their own original capital and to narrow its place in the field of production. Such a behaviour would be reasonable only if they occupy a monopolistic position in the sphere of production. Competitive capitalists would rather seek to apply their capitals productively as much as possible even if in vain."
14. I will also use, following Sekine (1986a, 541), such terms as "widening" accumulation and "deepening" accumulation as being synonymous respectively with accumulation with constant organic composition of capital and accumulation with rising organic composition of capital.
15. I dare not say that Marx completely neglected the absorption of labour-power. Indeed, Marx noticed this aspect In Volume I, saying as follows: "the course characteristic of modern industry, viz., a decennial cycle (interrupted by smaller oscillations), of periods of average activity, production at high pressure, crisis and stagnation, depends on the constant formation, the greater or less absorption, and the reformation

of the industrial reserve army or surplus-population. In their turn, the varying phases of the industrial cycle recruit the surplus-population, and become one of the most energetic agents of its reproduction" (Marx 1965, 632–3). However, if such a view were firmly established in Marx's theory of capital accumulation, he would not have given the so-called thesis of the absolute or relative impoverishment of the working-class.

16. I owe these considerations to Mawatari (1973c, 61–2).
17. Relating to this point, Hoshino (1977 and 1980) agreed with Kobayashi (1976 and 1979). Kobayashi's point is that a discrepancy between demand and supply does not arise in principle without credit.
18. Kawai (1957) and Furukawa (1959) were among the first orthodox Marxists in Japan who criticized Uno's treatment of the price in its relationship with wages.

6 The Social Contexts of Money Uses[1]

Makoto Maruyama

The purpose of this paper is to present a new frame of reference which might be useful for the study of money uses in empirically observable (rather than theoretically defined) societies. Both economics and political economy have studied money primarily from the point of view of the commodity-economy. Most economists today take it for granted that money functions mainly as a means of exchange, and assume that the existence of money *ipso facto* implies the presence of markets and trade. In other words, they presuppose the "catallactic triad" of money, market and trade to hold as a matter of course, when they raise the question of money uses.

The market-economic approach to money has been challenged, however, by economic anthropology. The latter claims that the catallactic triad did not automatically develop in all forms of economic life. Karl Polanyi (1968), who most distinctly adopts a substantive approach to economic studies, claims that interactions among human beings in their livelihood occur, as a rule, in such a way as to fit the forms of socio-economic integration, forms which may be supported by non-market structures.

In empirically observable societies in which the economy is embedded, money often appears separately from trade and markets. Even if trade and markets are present, the ways in which they relate to the use of money vary from one society to another. For example, the use of cowrie currency in the eighteenth-century Dahomey was confined to local markets inside the kingdom, and did not extend to foreign trade with outsiders. The stone money in the Yap Islands is employed even today for the purpose of clearing comumunal obligations among the natives, but is never used in trade with others. These monies are obviously different from the usual means of exchange which, according to Schurtz (1898, 6) "circulates from tribe to tribe and finally becomes a commodity that is generally accepted".

Thus the present paper, first of all, deals with what makes the market-economic approach to money distinct, making use of Marx's value-form theory (Marx, 1971a). Secondly, it illustrates the functions of money within the context of a purely capitalist society. Finally, it demonstrates the practical ways in which different types of money are used in different social contexts.

6.1 VALUE-FORM THEORY

When commodity exchanges (as opposed to use-value exchanges) occur, the existence of a market is already presupposed, in which traders appear as commodity-owners. They bring their commodities to the market with a view to exchanging them for others that they want. Commodity-owners are free from any ethno-cultural relationships which may socially bind them to others. They are presumed to act purely economically as isolated individuals who pursue their own interests in market transactions.

A "society" in which human interactions occur exclusively through markets would remain a stark utopia, unless labour-power were completely commoditised. When labour-power is converted into a commodity, workers have to sell their own labour-power in order to obtain the wherewithal to live. They become wholly isolated individuals in the sense that they have no relatives or friends with whom to reciprocate favours and obligations, in particular, by offering and receiving goods or services. In a market society, humans exist only as commodity-owners. By contrast, non-market societies, in which labour-power is not commoditised, makes people commodity-owners only when they have surplus products to dispose of. They come to a special marketplace where they exchange surplus products with aliens.

A commodity has two aspects: use-value and exchange-value. Use-value refers to that aspect of the commodity which serves the material want-satisfaction of the person who consumes it. Exchange-value is an expression of the value of the commodity in terms of the use-value of another commodity. A commodity always presupposes its owner, and the owner of the commodity is not interested in its use-value. He is interested only in its value, which expresses itself in an exchange-value. Whereas gifts and valuables, which are reciprocated among pre-modern people, strengthen the communal ties among them, commodities which their owners exchange among themselves render human relations among them at arm's length and indifferent.

Thus a society built on commodity exchanges requires a set of mercantile rules which are quite different from the rules based on communal interaction. Marx's theory of value-forms is particularly revealing with regard to the nature of commodity exchanges. In the following interpretation of it I owe much to the restatement of Marx's original idea by Uno (1980) and Sekine (1984).[2]

6.1.1 Simple Form of Value

When a commodity-owner goes to the market intending to exchange his commodity for another in possession of others, he must express his intention in a language which is understandable to other commodity-owners. This language must be in terms of value and use-value, the two elements of a commodity. Suppose the owner of commodity A wants another commodity B. He then announces that certain amount of A is worth the desired quantity of B. This announcement is put forward as an exchange proposal by the owner of A to the effect that a certain amount of it is available for a given quantity of B, and that he is ready to give away his commodity to anyone who agrees to the terms of proposed exchange.

Marx (1971a, 139) states this exchange proposal as "x commodity A $= y$ commodity B", where x and y are the quantities of commodities A and B respectively, and calls this the simple (or the first) form of value. He further states that A is in the relative form of value (meaning that it is for sale) and B in the equivalent form (meaning that it is desired). The simple form of value is also sometimes called "accidental", since it is only by accident that we find the interest of the owner of A in B.

6.1.2. Extended Form of Value

The owner of commodity A usually wants not only commodity B but also a variety of other commodities, so that his value-expression is extended as follows:

$$
\begin{aligned}
x \text{ commodity } \mathbf{A} &= y \text{ commodity } \mathbf{B}, \\
x' \text{ commodity } \mathbf{A} &= y' \text{ commodity } \mathbf{C} \\
x'' \text{ commodity } \mathbf{A} &= y'' \text{ commodity } \mathbf{D}, \\
\cdots\cdots\cdots \\
\rho \text{ commodity } \mathbf{A} &= q \text{ commodity } \mathbf{N}.
\end{aligned}
$$

This tabulation Marx calls the extended (or second) value-form (1971a, 154–155). It lists all exchange proposals by the owner of commodity A. In order for any of these proposals to be accepted, so that any exchange can actually take place, however, there must be another individual whose exchange proposal happens to be just the reverse of one of the **A**-owner's. At this stage, we are not yet sure if any exchange can actually take place in the market. We must, therefore, move to another form of value, which implies the possibility of general commodity exchanges. Marx calls such a form of value the general (or the third) form of value (1971a, 157).

6.1.3 General Form of Value

The general expression of value is stated as follows:

one unit of commodity **A** = x commodity **U**
one unit of commodity **B** = y commodity **U**
one unit of commodity **C** = z commodity **U**
. . . .
one unit of commodity **N** = ρ commodity **U**

What is apparent in this form is that all commodity-owners who do not themselves possess commodity U want to have it. They do not, however, want to get it because of its original (material) use-value, but only for the newly acquired (social) use-value of being a general purchasing-power in the market. The original use-value of commodity U has become insignificant for them. They seek to have it only because they want to purchase other commodities with U. In other words, they want U strictly as the means of purchasing other commodities. That is the reason why, in this form, only one unit of each commodity is offered for sale or placed in the relative value-form. Each commodity-owner wants as much of U as he/she can get per unit of his/her own commodity.

Marx claims that the general form of value can be obtained simply by reversing the extended form of value, i.e., by switching the commodity in the relative value-form for the commodities in the equivalent form. He does not explain how such a reversal is economically justified. In fact, there seems to be no good reason to explain such a procedure. A commodity-owner can always offer his own commodity for sale, i.e., put it in the relative form of value, but he cannot demand his own commodity, i.e., put it in the equivalent form. We must, therefore, justify the introduction of the general form of value in a different way.

In the expanded value-form of each commodity-owner a whole list of equivalent commodities appear. It is likely that some commodities are desired as equivalents by many commodity owners, while others are not. The commodities that appear most frequently as equivalents are the ones that are more widely demanded, and so can be sold relatively easily for other commodities. In fact, such commodities can serve as vehicles to buy other commodities. As commodity-owners acquire such commodities not for consumption but as means of indirect trade to achieve their ultimate end, a single commodity tends to emerge as the general equivalent. Once the market identifies the general equivalent U, the demand for it is no longer quantitatively limited. All commodity-owners want to have as much of it as possible for their commodities.

6.1.4 Money Form of Value

The general equivalent is a special commodity which never occupies the position of relative value-form. It is never offered for sale, but is always in demand, i.e., it always occupies the position of equivalent value-form. At this point, it has become money. The value-form theory shows how the exchange of commodities identifies without fail a commodity-money among themselves.[3]

6.2 FUNCTIONS OF MONEY

The commodity-money, the necessity of which has been demonstrated by the value-form theory, has the following functions: (1) the measure of value, (2) the means of circulation and (3) a store of purchasing-power (hoarding, the means of payment, etc.). These functions can be explained logically, i.e., without referring to any particular form of human interaction in society.

6.2.1 The Measure of Value

In the simple value form: "20 yards of linen = one coat", the linen-owner expresses the value of his commodity by offering 20 yards of it for the coat of his liking. This, however, is an exchange proposal, a subjective statement, in which the offer of 20 yards is tentative. If the market does not respond to his offer, he may have to give up more of his linen for the coat. Or if too many coat-owners are willing to accept his proposal, the linen-owner has to determine which one would settle with the least amount of his linen in exchange for the desired coat. In any case, he must revise his value-expression in light of market conditions.

The same holds for the general form of value: one unit of linen = p commodity U. In this case too the linen-owner is quoting the supply price of linen in money, tentatively. Only when many money-owners respond to this proposal, and actually buy linen with money for this price, is the value of linen measured. In other words, the value of linen is measured when its supply price, which its sellers quote, agrees with its demand price for which money-owners in the market are willing to buy. Money functions as the measure of value because it is a special commodity (which always occupies the equivalent form of value), that is to say, it is strictly the means of purchase and not an ordinary commodity which must be offered for sale.

6.2.2. The Means of Circulation

Money is the commodity which is generally acceptable, not for its material use-value, but for its being a social purchasing-power. It is this privileged fact that enables money to act as means of exchange.

In the market a great number of commodity exchanges take place in succession among different commodity-owners. For example, a tailor sells his coat and obtains money for it. With the money so obtained he buys a pair of shoes. In doing so, he lets the shoe-maker sell his shoes, giving him money. This money is further spent to buy a book, etc. In this way, the chain of commodity exchanges expands endlessly. In the meantime, money always mediates the circulation of commodities.

Although money is not demanded for its material use-value, but for the social use-value of mediating exchanges, it is originally a material use-value nevertheless like any other commodity. If this were not the case, it would not have been selected as the general equivalent from the whole class of commodities. It, therefore, makes sense to say that a thing which has no material use-value to speak of, such as paper money, cannot function as means of circulation, unless its use as money is dictated by fiat, i.e., from outside the market. Fiat money circulates only when its issuer (whether the national state or local government) is trusted by the commodity-owners. For example, in local markets in which local commodities are exchanged, paper money issued by the local authority can be used. But the scope of its circulation may be limited to the local region. A commodity money is needed in a long-distance trade in which traders from different communities deal with commodities of distant origins.

6.2.3 Hoarding

Money can also be withdrawn from the market as a store of value and be accumulated as mercantile wealth because "money is itself a commodity, an external object capable of becoming the private property of any individual" (Marx, 1971a, 229–30). Since money is the general equivalent, people can choose between using it in the present market, and using it in another market in the future. In the latter case, money is stored (hoarded) outside the market pending its return to another market.

In this case the distinction between commodity-money and paper-money is even clearer. A commodity-money, "itself [being] a commodity", can be held as a private property and can become the "absolutely social form of wealth" (Marx, 1971a, 229). Paper money which circulates by fiat, by contrast, loses its purchasing-power as soon as it leaves the sphere of commodity circulation. What I call paper money here is, of course, the one that circulates by fiat and does not include credit money issued by banks. I shall presently comment on the latter.

6.2.4 Means of Payment

A special variant of money as a store of value is means of payment which cancels a debtor-creditor relation previously formed. A debtor-creditor relation can arise from both market and non-market transactions. If it arises from market transactions, it is a relation between commodity-owners who are strangers to each other and come into contact with one another only by chance. They do not give credit to others as a personal favour. Instead, as Marx rightly states, "when the same transactions are continually repeated between the same persons, the conditions of sale are regulated according to the conditions of production" (1971a, 232).

Suppose that it takes three months for a cotton spinner to make a certain quantity of yarn. If the spinner goes to buy raw cotton from different farmers at random, he has to have enough money for the purchases. If, however, he buys cotton routinely from the same farmer, and is known to the farmer to be a good manufacturer, the farmer would trust him and would value him as an important customer. The farmer will then be ready to give him three-months credit, provided that the farmer has enough money to carry on his business even without receiving cash from the spinner during the credit period. In this case, the spinner will pay the farmer money as the means of payment in three months' time, which then cancels the debtor-creditor relation. Unlike the means of circulation, the means of payment does not mediate the process of commodity exchanges. Instead, it "brings [the process] to an end by emerging independently, as the absolute form of existence of exchange-value" (1971a, 234).

The above-mentioned debtor-creditor relation presumes a type of society in which market exchanges are accepted as an ordinary form of socio-economic integration. In such a society, people produce use-values primarily for sale, i.e., as commodities, rather than for their own consumption. As commodity producers, they must depend on markets to buy the wherewithal to live as well as the raw materials which they need for their commodity production. Markets in which people sell only their surplus products remain supplemental to the economic life of society. But in a society of commodity producers markets play a vital role in organising economic life, and the credit sale of commodities, especially of raw materials, tends to become routine. Accordingly, the need for means of payment increases. The latter is money set aside from active transactions and "hoarded", for the purpose of cancelling debt contracted in the past.

6.2.5 Credit Money

When a trader purchases a commodity on credit from another, the former issues a promissory note or bill of exchange. It obligates the payment of a certain sum of money from the debtor to the creditor on a stated date. To some extent the bill of exchange can circulate among traders. For example, the development of cotton trade creates interdependence not only between the farmer and the spinner, but also between the spinner and the weaver, and between the weaver and the tailor. For the farmer sells cotton to the spinner, who sells yarn to the weaver, who sells cloth to the tailor.

In such a chain of transactions, a trader can be simultaneously a debtor and a creditor of the same sum of money for the same period of time. For example, the spinner can give credit for three months to the weaver, and receives the latter's promissory note, which he can pass to the farmer with a suitable endorsement. A promissory note is a primitive form of credit money. It only circulates within a group of people who can make a chain of credit transactions, as commodities are sold on credit. But if there is no use-value connection between the two producers, they cannot develop a credit transaction or substitute a promissory note for cash payment. For example, a bottle maker may accept the promissory note of a brewer, but perhaps never the promissory note of a spinner.[4]

The restriction on the circulation of promissory notes can be overcome to some extent by the wholesaler who deals with commodities which belong to different industries. A large trading company can deal with commodities related to the cotton trade as well as commodities related to bottle-making. The wholesaler can also replace the promissory notes of his customers with his own commercial bill. For instance, the trading company can issue its promissory note to the bottle-maker, when buying bottles from him, and with the cash it does not have to use for the next three months, the company may buy the spinner's promissory note from the farmer. In this case, the bottle-maker is indirectly financing the spinner, and the promissory note of the trading company is mediating this.

The promissory note issued by a reputable wholesaler is more widely accepted by commodity producers as a substitute for means of circulation. But it is still a commercial bill so that it must be honoured on the date of expiry of the credit period. Banknotes which commercial banks issue are special promissory notes which have no date of expiry. They are honoured on sight by the issuing bank whenever they return to it. But since they have no date of expiry, they can circulate so long as conversion into cash is not contemplated by the trader who holds them.

The issuing bank has learned by practice that it needs to hold only a fractional cash reserve in order to meet the regular demand for conversion. The cash reserves of the banking system constitute, in effect, society's means of payment, on the basis of which a much larger sum of credit money (banknotes) can be created. Credit money backed by cash reserves serves as means of circulation of commodities.

6.3 MONEY USES IN EMPIRICALLY OBSERVABLE SOCIETIES

In the previous section, the nature of money in a purely capitalist society has been reviewed. A commodity money is automatically generated from commodity exchanges. Such money may be called "all-purpose money" or more preferably "multi-purpose money". However, the two kinds of paper money which we referred to above may, with some qualifications, be deemed as special-purpose money. Paper money issued by a political authority can function as means of circulation only within the scope of that authority and not beyond its boundary. The circulation of credit money (banknotes or demand deposits) is also limited to the sphere within which the reputation of the issuing bank is well established among the traders. In any case, we have so far treated the question of money uses theoretically, i.e., from a logical point of view. In the present section, however, I intend to consider the question of money uses from an empirical point of view.

6.3.1 Money Uses in Modern Societies

Most societies in the modern world utilise central banknotes exclusively as currency. These banknotes combine the properties of both fiat money and credit money. The central banks are national institutions and the banknotes that they issue circulate within the nation as legal tender. These banknotes are not convertible into commodity money. The central banks regulate the money supply including the supply of notes by buying and selling national bonds and treasury bills. The currency of a modern nation is, therefore, partly credit money and partly fiat money, and itself functions as non-convertible multi-purpose money.

The evolution of such modern currency after the fall of the international gold standard is a political event, and cannot be theoretically explained by the logic of commodity exchanges alone. A highly market-oriented economist like F. A. Hayek (1976) still claims that money should be denationalised, and that private commercial banks should be permitted to issue their banknotes freely. Being an arch liberal, he presumably wants to separate the economic from the political even today when the two are so deeply interwoven.

6.3.2 Money Uses in Non-Market Societies

A non-market society is one which has no market system. This, of course, does not mean that the society has no markets. It may, on the contrary, have many markets for use-values or products of labour. But it does not have the organised markets for labour-power, land, and money. According to Polanyi, the markets for these three "fictitious commodities" are prerequisites for the self-regulating system of markets.

In a non-market society, money appears primarily as special purpose money. Some monies are used solely as means of exchange in the local markets for foods and other immediate necessities. These monies are often represented by objects which have little economic value, such as cowry shells. In substance, they are akin to the paper money in a commodity producing society. For international or inter-regional trade, however, a commodity money such as gold is used. As for the means of payment to clear social obligations, economic or non-economic, another type of money (often not easily portable) is employed. It has to have some social value, but certainly need not be a commodity money. Whatever is valued in society by tradition, for economic or non-economic reasons, can be used as the means of payment.

6.3.3 Local Currencies

In 6.3.2 above we discussed fiat paper money which may be issued by local authorities. Such a paper currency can circulate within the sphere of influence of the authorities, namely, in local communities. In such communities, the users of the locally issued currency need not be commodity producers but "people in their livelihood". The exchange of goods and services among local people can include reciprocation of gifts and courtesies. In such communities, the local currency can be used as means of reciprocity. The word "reciprocity" is chosen here in preference to "exchange" because of its emphasis on the direct interaction among the people of the community, rather than on the arm's length transactions among commodity producers.

Theoretically as well as practically, the local currency can be issued not only by the community authorities but also by the community people themselves. A good example of the latter is the Green Dollar of LETS (an acronym for Local Exchange/Employment and Trading System) in Canada and in other countries (Ekins, 1986). The green dollar functions as if it were a deposit currency subject to chequing. It is issued by the members of LETS, when they purchase the products or the services of other members. The difference between the usual deposit currency and the green dollar lies in the fact that unlike the former the latter is not convertible into cash. A green dollar always remains within the LETS. It circulates among the accounts of LETS members but never drains away from their community.

The use of local currencies does not contradict the use of national or international money. It serves to diversify markets, and stimulates the economic activities of the local people.[5]

6.4 CONCLUSION

Walter C. Neale (1976) writes that "[in] thinking about money it is best to think in the plural-about monies. It is also best to think, not about money in the abstract, as having some universal role in social and economic life, but to think about how each money fits into the social, political, and economic structures of the society which used that money". This paper has tried to confirm this proposition by suggesting the one-sidedness of the market-economic concept of money. It has especially tried to shed light on the fact that the socio-economic context of the special-purpose monies cannot be explained logically from the viewpoint of the commodity-owners in a purely market society.

Notes and References

1. This paper is dedicated to Professor Thomas T. Sekine, one of my mentors. He has encouraged me to undertake interdisciplinary studies involving both political economy and economic anthropology. I am also grateful to Professors Robert Albritton and Colin A. M. Duncan for their valuable comments on an earlier version of this manuscript. They are incorporated in the present version.
2. Orthodox Marxists have emphasised that the labour theory of value is essential to the understanding of exchange-value. Marx himself states that exchange-value reflects "abstract general labour" embodied in a commodity, whereas its use-vale is created by "concrete useful labour". An argument of this sort merely repeats the contention of the Ricardian labour theory of value. What Marx actually does in his demonstration of the labour theory of value is to show that the capitalist mode of production rests on commodity exchanges in general. In this argument, however, Marx does not sufficiently emphasise his other important observation that commodity exchanges begin at the point were communities meet. Uno (1980) separates the labour theory of value from the theory of commodity exchanges in order to bring out the significance of Marx's value-form theory. Making much of value-form theory does not, however, mean to deny the validity of the labour theory of value. Uno claims that exchange value obtains its objective base when labour-power is fully commoditised and thus all necessities of life must be bought back by wage-workers. It is the value of commoditised labour-power that provides the key to understand the meaning of the labour theory of value. On the relationship between the value-form theory and the labour theory of value see Sekine

(1984), Itoh (1980), and Albritton (1986). The present exposition of the value-form theory mainly follows Sekine's treatment (1984, 128ff).

3. Marx writes that "the specific kind of commodity with whose natural form the equivalent form is socially interwoven now becomes the money commodity, or serves as money" (1971a, 162). Logically, however, it is hard to determine which commodity finally becomes the money commodity. The reason for choosing gold as an example of the commodity money is historically given. "Among the commodities . . . there is one in particular which has historically conquered this advantageous position: gold" (1971a, 162). Also, "the form of direct universal exchangeability, in other words the universal equivalent form, has now by social custom finally become entwined with the specific natural form of the commodity gold" (1971a, 162).
4. See Sekine (1986,399ff) for further details on trade and commercial credit.
5. In explaining the viability of local currency, I wrote: "the use of local currency can activate the community life of local people in accordance with their vernacular culture" (Maruyama,1988, 74).

Part II

Philosophical Economics

7 Dialectics and Economic Theory

John R. Bell

Commodity-economic logic does not arise inevitably in the history of human societies. While this logic necessarily prevails to the degree that a society-wide capitalist market dominates economic life, it is a contingent historical development which transforms not only material products but also human labour-power, the ultimate source of productivity, into commodities.

This could never have taken place unless the direct producers were first separated from the means of production, and unless traditional economic relations were gradually eroded by commodity-economic relations. This process occurred slowly over a period of centuries and could, in principle, have been reversed or halted. But following the industrial revolution in late 18th century England, capital was finally able to commoditize the labour-power it required, while managing the production of use-values as commodities and securing access to land and its resources. The capitalist market could then regulate price, profit and wage levels throughout the economy such that the principal classes of capitalists and workers received the incomes and goods required to ensure the reproduction of both labour-power and capital. Social reproduction was thus no longer guaranteed by the direct human relations of dominance and subservience which had characterized earlier societies but by the anonymous, impersonal and reified market.

It is not a trivial or simple task to attempt to determine what it was about the nature of the society-wide, competitive, capitalist market that allowed it to successfully reproduce material economic life, including the material requirements of the two major classes, when the state increasingly adopted non-interventionist economic and social policies as it did in liberal England. I believe that this was the task Marx set for himself in the three volumes of *Capital*. It was Marx's intuition that capitalism's survival indicated that the competitive market must operate according to a rigorous logic. Unfortunately, Marx died before he completed his study of that logic in *Capital*.

Liberal capitalism, the purest form of capitalism ever to appear in history, held sway in mid-19th century England when Marx was writing *Capital.* That made Marx's task of attempting to comprehend capitalism's laws of motion, or logic, easier than it was for earlier theorists of capital such as Adam Smith. It might be supposed that any attempt to understand the laws of motion of liberal capitalism from our vantage point today would be a more difficult task than it was for Marx because that society has passed into history. Beginning in the late 19th century, during the stage of capitalism which has become to be known as the period of imperialism, capitalism began to move away from regulation of the economy by the competitive market because the production of iron and steel products with new, heavier, and more expensive technologies led to the rise of giant, oligopolistic corporations and increasing state intervention. This reversed the early history of capitalism which was characterized by a movement in the direction of the automatic regulation of economic life by the market.

Even if this sort of society were still with us, we could hardly study it in the laboratory, as is sometimes possible when one is investigating certain objects in the natural sciences. However, if liberal market capitalism was indeed largely governed by a commodity-economic logic it should, in principle, be possible for us to reconstruct that logic in thought, provided that we can find the right starting point to activate that logic and if we then allow that logic to unfold until it reaches a self-imposed closure. Kozo Uno, the Japanese Marxian political economist, did just that. He believed that Marx had indeed found the right starting point to expose the inner logic governing the operation of the society-wide market in liberal capitalism. He recognized, however, that Marx did not always adhere strictly to the logical exposition which makes it possible for the theorist to reproduce capital's laws of motion in thought. Marx had in fact made many methodological errors in his attempt to copy, retrace or reproduce in thought the method which capital employs to regulate the material economic life of a society. By adhering much more rigourously to the dialectical methodology employed by Marx than even Marx himself, Uno was able to largely complete and correct Marx's explanation of the laws of operation (or logic) of capitalism. Uno's *Principles of Political Economy* (1980) has been translated into English by Thomas T. Sekine. Sekine has also made a major theoretical contribution himself by making explicit the dialectical methodology employed falteringly by Marx and more consistently by Uno, thus demonstrating that Uno's account of the logic of capitalism was not just another one-sided Marxian interpretation of the nature of capitalism, but a complete definition (specification) of capitalism as "what it is". Sekine has also introduced significant mathematical and methodological refinements into the theory of capitalism. Sekine's contributions are best viewed in the context of his major work *The Dialectic of Capital* (1984, 1986).

The Uno/Sekine dialectical theory allows us to achieve an understanding of how the society-wide commodity-market was able to reproduce the material economic life of capitalist society, including the material requisites necessary for the reproduction of the capital/labour social relation in the absence of significant state economic intervention. What was it about capitalism, as opposed to feudalism or any other economic system, which makes an objective account of its operation both necessary and possible? For one thing, capitalism systematically reifies or objectifies economic relations as impersonal, anonymous commodity-relations. To the degree that workers and capitalists tolerate the existence of a society-wide market for material commodities and commoditized labour-power, and do not demand too many heavy or complex use-values which small competitive firms cannot produce, the society-wide competitive market is able to direct human economic activities in such a way that capitalism is reproduced over time, without significant state or community-based economic policy intervention. Thus, the market, through its commodity-economic logic and not human agency, manages the greater part of economic life.

It is, however, quite difficult to determine just what this logic is for a variety of reasons. To begin with no capitalist society has ever been so completely dominated by this logic that a straightforward empirical/historical investigation would ever allow one to lay bare the logic of the system with perfect clarity. There has always been some collective resistance to the market which impedes its operation in even the most competitive capitalist societies. Moreover, the heavier and the more complex the use-values are that capitalist society demands the more difficult it is for capital to autonomously manage material economic life. Thus, capital's laws of motion tend to be impeded in their operation to the degree that capitalist society requires the production of iron and steel products as was the case in the imperialist stage of capitalism. It was fortuitous for capitalism's continued development then that in the first two stages of capitalist society's historical development the dominant use-values demanded were such light use-values as woollen goods in the stage of mercantilism and cotton goods in the stage of liberalism.

Capitalism's unique commodity-economic laws of operation and the general norms of material economic life which must prevail in all viable societies are very difficult to distinguish, especially in a society in which the state becomes progressively less involved in economic policy formulation and permits the market to regulate material economic life. The intermingling of the commodity-economic and the material-economic confuses not only ordinary citizens but also neoclassical economists as well. In the absence of a clear grasp of the logic which governs their economic life in capitalist society, citizens of that society may accept the liberal dogma that they are unconditionally free as economic agents or, because they are unable to separate the specifically capitalist economic laws which are historically transient from the general norms of economic life which must prevail in all viable historical societies, they may equally well conclude that the economic laws of capitalism are universal and unchanging features of the economic life of any society.[1]

Finally, it is also possible to confuse or conflate the logic which governs a mature capitalism with the historical process by which capitalism achieved that maturity. A theory of capitalism cannot simultaneously explain the operation of a mature capitalist economy, which can only begin to operate once certain preconditions have been satisfied, so that capital is able to reproduce itself primarily by commodity-economic means and, at the same time, account for the system's historical rise and fall. It should be obvious that not all of the preconditions which must be present in order to ensure capitalism's reproduction primarily by its own commodity-economic logic are going to be present when capitalism is just beginning to emerge as an historical institution nor, indeed, when the system is in its death throes and the commodity-economic logic of the market is losing its grasp over the labour-and-production process common to all societies. Tracing the evolution or devolution of the preconditions which make an all-embracing capitalist market a viable way to organize economic life is, therefore, quite a different task than that of explaining how capital's logic manages to ensure the reproduction of economic life when those preconditions are in place. There is no justification, then, for viewing the logic of capital as a purified history. The logic focuses on the necessary features of capitalism which must be present to the degree that capitalism really exists in history, whereas the historian who wishes to explain the system's rise and fall must be familiar not only with what must necessarily be present in any viable capitalist society. [S]He must also be able to explain which contingent historical developments permitted the logic of capital to organize a greater or lesser part of economic life as capitalism matured, and which contingent historical developments later undermined the system's viability.[2]

I have said that the Uno/Sekine theory of the logic of capital provides us with an objective knowledge of the laws of motion by means of which a capitalist market economy regulates itself. I must qualify this statement by again reminding the reader that this is true only to the degree that the competitive market really does regulate the economic life of a capitalist society. The Uno theory does not make the claim that capital with all its cunning ever had the power to resist and overcome any and all attempts by the working class or any other classes or groups to overthrow the system. The Uno theory acknowledges that to the degree that state intervention, oligopolistic corporate concentration, class struggle, and use-value resistance interfere with the operation of the society-wide competitive capitalist market, an objective account of how the system operates will not be possible. However, it is also true that to the degree that the competitive market does not regulate the economic life of a society, then that society will be less capitalist at least in the sense that the term has come to be understood by most of the friends and foes of capitalism alike. Thus, a society in which the competitive market has lost its grasp over the greater part of economic life would no longer be capitalist.

No society has ever allowed the competitive market to dominate all of its economic life but up until the last quarter of the 19th century, it is undeniable that there was a tendency in the leading industrial countries such as England for the market to increasingly regulate economic life. The fact that the liberal utopia of a purely capitalist society was never completely realized in history should not deter us from using the Uno theory of the laws of motion of a market-based capitalist economy to simultaneously expose the limits of that liberal utopia and to lay bare the laws of motion which really do govern capitalist societies in history to the degree that all classes of people in these societies have allowed the market to regulate their economic activity.

Because the logic of capital was always checked to a degree even in liberal capitalism's heyday, Uno and Sekine recognize the necessity to theorize a purely capitalist society in which all production is organized as the production of commodities by means of commoditized labour-power and commoditized material inputs in order to view these laws with perfect clarity. The theory assumes that the state does not interfere with the free movement of commodities, labour-power, or capital and advances no active economic policy. It assumes as well that competition, not oligopoly or monopoly, prevails everywhere in the economy. The fact that capitalism survived in history with progressively less state economic policy intervention in the commodity-economy (until the late 19th century) suggests that this economy governed itself by such a logic, even if that logic only operated as a tendency in history due to an omnipresent, though varying, resistance to the market's operation.

The Uno theory of pure capitalism should not be confused with the ideal types favoured by liberal social scientists because it is not constructed by a one-sided and subjective selection from innumerable conflicting tendencies of just those tendencies or characteristics which the theorist arbitrarily deems to be of greatest importance. By contrast the Uno/Sekine theory does not leave anything essential about capitalist economy's logic unexplained. It can do this because capitalism is a reified society which has the power to reproduce itself by the operation of its own commodity-economic logic in the absence of significant community or state planning. It is possible for a theorist to reactivate in theory capitalism's power of self-regulation which hitherto existed as an objective reality. When the theorist performs this operation [s]he ends up retracing the path or steps by which capital went about organizing economic life, and so allows capital to reveal its laws of operation in their entirety. To the best of my knowledge no other theory of a social institution can make this claim.[3]

7.2 AN INTRODUCTION TO THE DIALECTIC OF CAPITAL

The logic by which capitalism regulates itself is dialectical in the Hegelian sense. For the dialectic the truth is the whole. To say that we comprehend capitalism is to say that we know the "whole truth" about capitalism (namely, its inner logic) and that we have maintained no unverified hypothesis or conjecture about that system as a subject/object (i.e., theoretical object[4]). If we have truly achieved complete knowledge of the system's inner laws of motion (i.e., how it is "programmed" to operate), then our theoretical explanation of capitalism as an object of investigation will be identical with the logical self-exposition of capital as the subject. The construction of a purely capitalist society as a dialectical system in thought is, therefore, justified and is possible only if our object of investigation (i.e., capitalism which occurs in history) has an innate tendency to reveal itself as a self-contained and self-determining entity governed by a teleological logic.

In order to grasp a subject-matter totally, the dialectic proceeds step by step from the abstract to the concrete. The concrete in this context should not be viewed as the real or empirical concrete. At the abstract starting point of the theory, the subject/object and its conceptualization are as yet wholly unspecified and empty because the subject/object cannot be grasped in all its synthetic concreteness by immediate conception. Only when the theory has been fully developed, does the subject/object become concrete-synthetic in the sense that it is fully specified with all determinations. The dialectic always presupposes the fully specified totality as its subject-matter, but does not exhibit all of its concrete features (specifications) at once. Instead, the initially empty and abstract whole is gradually enriched with specifications in increasing detail, until all of its logical components (or necessary inner connections) are exposed. Every time a new specification is introduced, the level of abstraction (synthesis) is changed. At each level of abstraction (synthesis), the dialectic recognizes a "contradiction", the synthesis of which amounts to a new specification of the totality, enabling an advance to a more synthetic level, until, finally, the dialectical circle reaches its closure.

At this point, there are no more contradictions left to impel the logic forward, for indeed the subject-matter needs no further specification. The dialectic, in other words, returns to the starting point of its exposition in possession of all the "concrete" details which are necessary to fill the original emptiness and abstractness of the concept. Thus, dialectical concreteness does not mean "empirical" but rather "synthetic" concreteness. The dialectic ends with a fully synthesized totality which Hegel refers to as the "absolute idea". It is at this point that the subject-matter completely and, therefore, absolutely, reveals all its inner connections.

Hegel would no doubt have maintained that the material world could not contain a dialectical (i.e., self-revealing and self-determining) object. However, there exists in fact, at least one such object, and that is capitalism. In the materialist dialectic, developed by Marx and Uno, capital plays a role quite similar to that occupied by the Absolute in Hegel's *Science of Logic* (1969). The logic of capitalism tends to synthesize itself into a closed dialectical system, which a theorist can comprehend only by reproducing in thought that process of self-synthesis inherent in historical capitalism. Indeed, this inner logic could never be arbitrarily constructed in one's mind and imposed upon the external world. All such attempts have produced one-sided definitions of capitalism which are never logically complete. If a natural or historical object of investigation is not dialectically self-managed, then the decision to view that object as a logically synthesized totality can only be subjective and arbitrary. In this respect, historical "dialectics" favoured by Western Marxism are in no way superior to that of the Engelsian, Leninist-Stalinist dialectics of nature. On the other hand, the dialectic of capital is genuinely dialectical because the reified capitalist market has the potential to become a self-contained, self-determined, self-expanding and self-revealing dialectical totality, once a society permits capital to produce its key use-values as commodities.

To say that capitalism is self-contained is to say that it is an independent and internally coherent totality which is unified by a set of necessary inner connections. An attempt will be made within the context of this brief paper to demonstrate that all the categories of capitalism are internally related or necessarily connected. Because it contains within itself the means to ensure its own reproduction, capitalism is unique among economic systems. In this respect, capitalism is permitted to reify economic relations as commodity-relations, it does not have to rely on extra-economic, political intervention to guarantee its continued existence, as did pre-capitalist societies. During its heyday liberal market capitalism demonstrated a capacity not only to reproduce itself but also to expand to manage an ever greater portion of material economic life by undermining traditional economic relations. A completely reified (or purely) capitalist society is one in which all production is the production of commodities and all inputs in the production-process, including labour-power, are commoditized. In such a society economic life would be completely governed by the self-expansion of value (and of capital), while competition amongst capitals would ensure that the system was necessarily expansive. Although historical capitalism never achieved this purity due to use-value and collective human obstacles placed in its path, it increasingly displayed this capacity for self-expansion until late in the 19th century.

As capitalism matures, it also becomes increasingly self-revealing at least in the sense that, as the commodity-economic management of technical and material economic life comes to require progressively less extra-market intervention to guarantee the social reproduction, the market and its logic operate with greater autonomy and force and with less extra-market entanglements. As has already been observed, however, the logic never organizes all of economic life due to use-value and human resistance to its operation. Moreover, even in the absence of such resistance and interference, the market-economic and material-economic aspects of economic life under capitalism continue to appear superimposed one upon the other. Thus, even through the transparency referred to is real enough from one point of view, the task of the theorist of capitalism is still not an easy one.

I wish to repeat that the movement from the abstract to the concrete in a dialectical argument is not, as often understood, a movement from the abstract-general to the empirical concrete. It is a movement from insufficiently specified to more adequately determined categories or concepts. Thus, in the dialectic of capital, the argument begins with the commodity-form which represents capitalism immediately, i.e., prior to any further mediation (i.e., specification or determination). The dialectic then moves step by step to more synthetic categories, until it reaches the form of interest-bearing capital, which indicates the conversion of capital back into a commodity, to conclude its "spiral" movement.[5] Since the real-concrete has a separate existence from the concrete-in-thought, the sequence of categories in political economy need not correspond to the sequence of events in history. The dialectic reproduces the concreteness of its object of study not as a history but rather in terms of what is logically prior when it does exist. In the Doctrine of Circulation which opens the dialectic, the development of the theory closely parallels the historical sequence of events, but, later in the Doctrines of Production and Distribution, the logical order of the dialectic frequently reverses the historical order. Similarly, even though "money" is a more abstract and simple category than "cooperation", and even though money appears quite early in history, there have been societies which developed a kind of complex division of labour based on cooperation and yet did not develop money. Hence, the theoretically more concrete category can exist historically before the more abstract.

It can also be argued that money only reaches its full development in the generalized commodity production characteristic of capitalist society.[6] According to Marx (1973, 103) "this very simple category . . . makes a historic appearance in its full intensity only in the most developed conditions of society". Thus while it is true that money may "exist" in an embryonic form from very early in history, it only becomes a fully developed economic form in modern capitalism. The same considerations apply to the category "labour". Since people have always laboured, labour might well be the starting point for a political economy which follows the sequence of history. The problem with that approach, however, is that the notion of "labour as such" (i.e., as abstract labour or labour-in-general) only arises with the development of a fully developed market capitalism, which is built on the foundations of commoditized labour-power, and a very advanced division of labour.

As Marx argues:

> Indifference towards any specific kind of labour presupposes a very developed totality and real kinds of labour, of which no single one is any longer predominant. As a rule, the most general abstractions arise only in the midst of the richest possible concrete development, where one thing appears as common to many, to all. Then it ceases to be thinkable in a particular form alone. On the other side, this abstraction of labour as such is not merely the mental product of a concrete totality of labours. Indifference towards specific labours corresponds to a form of society in which individuals can with ease transfer from one labour to another, and where the specific kind is a matter of chance to them, hence of indifference. Not only the category labour, but labour in reality has here become the means of creating wealth in general, and has ceased to be organically linked with particular individuals in any specific form (1973, 103–4).

Thus, labour may be as old as humanity itself but, before the modern concept of "abstract labour" can be objectively or scientifically grounded, one requires the actual development of the capitalist labour market and "indifference towards specific labours." In other words, although productive activity is essential in all societies, it is the chrematistic from of capital that simplifies productive labour to the maximum degree compatible with the prevailing level of technology, and in so doing, establishes the labour theory of value both as a scientific concept and as the organizing principle of the commodity-economy.

7.3 THE DOCTRINES OF CIRCULATION AND OF BEING

Let us now explore the close correspondence which exists between the dialectic of capital and Hegel's dialectic of categories. Hegel's *Logic* consists of the three Doctrines of Being, Essence and the Notion.

Similarly, Uno's *Principles* and Sekine's *Dialectic of Capital* are divided into the Doctrines of Circulation, Production, and Distribution. The parallel between the logic of a purely capitalist society and Hegel's logic of Absolute Reason has been schematized by Sekine as follows:

Dialectic of Capital		Hegel's *Logic*	
I.	The Doctrine of Circulation	I.	The Doctrine of Being
A.	The Commodity-form	A.	Quality
B.	The Money-form	B.	Quantity
C.	The Capital-form	C.	Measure
II.	The Doctrine of Production	II.	The Doctrine of Essence
A.	The Production-Process of Capital	A.	Ground
B.	The Circulation-Process of Capital	B.	Appearance
C.	The Reproduction-Process of Capital	C.	Actuality
III.	The Doctrine of Distribution	III.	The Doctrine of Notion
A.	The Theory of Profit	A.	The Subjective Notion
B.	The Theory of Rent	B.	The Objective Notion
C.	The Theory of Interest	C.	The Idea

Source: Albritton, 1986, p. 187.

Sekine and Dunne have explored many of the parallels in the two dialectics[7]. I have relied extensively on their work in order to develop a short summary of the argument of the dialectic of capital (DC) together with a brief and necessarily impressionistic introductory discussion of the parallels with Hegel's *Logic* (HL). Since I intend only to provide a conspectus of DC while emphasising its parallels with the structure of HL for illustrative purposes, I have taken the liberty of treating some parallels in much greater detail than others and do not pretend to offer either an exhaustive or balanced account of the correspondence. If I succeed in arousing the reader's interest in this topic I will feel that I have accomplished my objective.

Both DC and HL begins with the most abstract, empty and implicit category which represents the subject-matter. In HL that category is the "pure being" of the Absolute Reason, and in DC it is "value" which is the most abstract representation of the presence of capitalism. Value, which is the simplest definition of capital, follows the "logic of transition" to pass over from one form to another, e.g., from the commodity-form to the money-form and, finally to the capital-form in the Doctrine of Circulation. Similarly, in HL "being" as the most universal, abstract, and empty category passes over to "nothing" and then to "becoming" in the Doctrine of Being. In both cases the first doctrine studies the mode of existence (or operating principles) of the subject/object without reference to its substantive content or inner determinations.

Although the dialectic of capital does not make it immediately explicit, the theory always presupposes a fully developed capitalist society in which all use-values are produced for a society-wide, competitive market by capital which employs commoditized labour-power and material inputs purchased as commodities in that market. The circulation-forms of commodity, money and capital presented in the first doctrine, are, therefore, implicitly capitalist categories and should not be confused with their pre-capitalist forerunners, even though the latter may indeed share some properties of the former. Just as "pure being" in the first instance appears in HL as the initial concept or starting point and is only later shown to have been the first glimpse of the Absolute Idea, so is the category of value initially introduced as the defining characteristic of the commodity and is only later comprehended to be capitalism itself in its immediate presence.

Indeed capitalism initially presents itself as an immense collection of commodities which have been offered for sale in the market place. The value of a commodity represents its social worth in relation to all other commodities. This comparability must stem from something other than the physical and heterogeneous properties of its use-value. At this level of abstraction in the dialectic of capital, however, we only need to take cognisance of the primacy of the commodity's "social" being, without further elaborating on the concrete nature of that social property. Thus, as value, a commodity constitutes a fraction of the homogeneous mass of society's abstract-general, mercantile wealth, whereas, as a use-value, it is a particular, isolated, individual sample from the heterogeneous aggregate of society's material wealth. When a commodity suppresses its own use-value in order to express its value in the use-value of another commodity it takes the form of exchange value.

The contradiction between value and use-value echoes the contradiction between pure, abstract and featureless "being" and "nothing" in HL. To the triad of "value, use-value, and exchange-value" in DC which sets the stage for value-form theory corresponds the triad of "being, nought and becoming" in HL which prepares the way for the logic of determinate being (Dasein).

The commodity-owner (seller) must be viewed as an implicit merchant (capitalist). He selects a determinate quantity of the use-value of another commodity which he requires for his personal/productive consumption in order to express the value of his own commodity. But his value expression, or pricing of its commodity, implies no actual exchange of commodities. The two commodities, his own for sale and the other that he desires cannot be exchanged without the mediation of money, which we have not as yet theoretically evolved. We are here concerned only with an "exchange proposal" in which the commodity for sale exists physically, but the commodity desired is only ideal, in the sense of existing in the mind of the proposer of the exchange.

For example, if **A** is the commodity for sale and **B** the desired commodity, the seller of **A** may express its value in the use-value of commodity **B** which he needs or desires in the following manner:

$$x \text{ amount of } \mathbf{A} = y \text{ amount of } \mathbf{B}.$$

This is called the *elementary or simple value-form*. In this exchange proposal, **A** (the commodity for sale) stands in the position of "relative value-form", and **B** (the desired commodity) stands in the position of "equivalent value-form". Despite the equality sign the relation of the two commodities is asymmetrical. The equation cannot be reversed, since no actual barter is taking place and since **A**'s exchange proposal does not imply the existence of **B**'s counter exchange proposal. To propose the exchange of **A** for **B** is to negate **A**'s own use-value and to adopt the use-value of the "other" commodity, in this case **B**, as the reflector of its value. In the logic of "determinate being" the Hegelian dialectic also specifies, determines or delimits the original pure being in terms of "others". Indeed, to say that "x amount of **A** is worth y amount of **B**" is to say that "x amount of **A** is not worth y' amount of **B**, or not worth y amount of **B'**." It is to specify the value of **A** by delimiting it with a finite. In Hegel, however, the finite limit is "alterable". In other words, the value of **A** can as well be expressed by:

$$x' \text{ amount of } \mathbf{A} = z \text{ amount of } \mathbf{C},$$
$$x'' \text{ amount of } \mathbf{A} = u \text{ amount of } \mathbf{D},$$
$$x''' \text{ amount of } \mathbf{A} = v \text{ amount of } \mathbf{E},$$
$$\ldots\ldots \text{ etc.}$$

This tabulation is called the *expanded form of value*. While it is a broader social expression of value than the simple value form, this form in which the value of **A** is expressed "alterably" by many finite limits can degenerate into what Hegel calls bad infinity. We can conceive of an infinite number of exchange proposals, each of which is an expression of the value of **A**, but none of which is either final or conclusive.

In the Hegelian system, coherence resides only in an internally related system of finite beings which are simultaneously subject and object. In such a system any particular finite is determined by the totality of all other finites, but simultaneously determines all the others. This is the system (rather than mere enumeration or string) of finites, or the True Infinite, in which each finite is both active and passive, i.e., determining and determined. Once again the logic of the Hegelian system points the way to the achievement of a more adequate expression of the value, or social relation to the commodity, in the community of traders. If a commodity trading system is to be an example of a Hegelian True Infinite, that which measures the value of any particular commodity cannot be some other commodity subjectively selected by the commodity-owner himself, but must instead be decided impersonally and objectively by the combined demand of all other traders, who possess a universal equivalent which embodies their combined demand. Any commodity-owner would then be a passive seller, the value of whose commodity is determined by the impersonal forces of the market just as the totality of finites or True Infinite determines any particular finite in Hegel's *Logic*. The first step in the direction of the True Infinite is for the trading community to overcome the limitations and the arbitariness of the expanded value-form. No longer will each commodity-owner make his private exchange proposals in terms of subjectively selected equivalents (as finite limits). Rather all commodity-owners will be constrained to make their exchange proposals in terms of one universally desired use-value which has thus attained the status of a social value-reflector or general equivalent.

Because the owners of the equivalent commodities found in the expanded form of value of, say, commodity-owner **A** likewise evaluate their commodity in terms of a similar series of equivalents, there is bound to emerge one or several commodities which is/are most commonly desired in the value expressions in the trading system. At first one or several commodities are recognized as being commonly desired for its/their use-value properties by commodity owners for their personal consumption. Once commodity owners recognize that many are projecting the value of their commodities in the value of the same equivalent(s) they begin to demand the commonly chosen equivalent(s) not primarily for their personal use, but because they know that such an equivalent will allow them to obtain any commodity they desire and not just these commodities whose owners also desire their commodities. Thus they list in their exchange proposals only one or several commodities which have the abstract social or mercantile property of immediate and universal exchangeability, and they endeavour to acquire as much of the said equivalent(s) as they can. The existence of several commodity-exchange systems within a trading community must be a transitory phenomenon. It is inevitable that one commodity will become excluded from the ranks of the commonly desired commodities to become the general equivalent. The general equivalent is that particular commodity whose "abstract-social" use-value of being immediately and universally exchangeable by all commodity-owners prevails over its natural use-value of being useful for consumption in one way or another.

No doubt one commodity will have physical properties which will make it a more attractive choice for traders to use as the socially chosen general equivalent; however, the logic cannot determine which commodity a specific community will choose. The development of the general equivalent marks the sublation of the use-value restriction in the expression of value because, commodity-owners no longer need subjectively or privately express the value of their commodity in terms of the equivalent use-values they desire for consumption. They achieve their desires indirectly by expressing value solely in terms of the commodity which has attained social recognition and as such embodies universal and immediate exchangeability. All commodity-owners now express the value or social worth of each other's commodities through this general equivalent. Each commodity is now socially related to all others in a interconnected trading community. When all owners express the value of their particular commodity in terms of a certain amount of the general equivalent, the *general form of value* is said to arise.

The *money-form of value* arises when gold or some other precious metal begins to act as the exclusive reflector of value. A precious metal lends itself to this task because it can be divided/fused without losing its homogeneous quality. Metallic money allows the diverse values of commodities to find a physical expression. Each commodity-owner expresses the unit value of his commodity in a given quantity of metallic money. The aggregate supply of a given commodity can then be subjectively evaluated in terms of the aggregate amount of gold desired. The amount of gold so required becomes its supply price. When all commodities are so priced, the *price form of value* emerges.

As suppliers of commodities merchants are primarily sellers who passively wait upon purchasers of their product. However, all commodity-owners must have sold their commodities in the past and are thus in possession of metallic money. Hence, as money-owners, merchants also represent the market demand for their own commodities. Their individual money demands can now also be aggregated in order to express the impersonal social demand for society's supply of all commodities.

Only as the concrete unity of the antagonistic forces of supply and demand can the system of commodity exchanges be established in such a way that negation falls completely within the self-determining totality of the market. This, of course, does not mean that the determinate, particular beings of commodities and commodity-sellers have vanished. Rather, they have lost their character of "self-subsistence". Only the whole is self-subsistent and only as a member of the whole is the finite being non-self-contradictorily defined.

The power of attraction which resolved the "diversity and externality" of the ones into the homogeneous "one One" in Hegel's *Logic* corresponds to the social integrating power of demand which subordinates merchant activity to the needs of the trading community. And just as the two Hegelian processes of Repulsion and Attraction, though distinct, formed two aspects of a single process of Being-for-Self, the sellers as representatives of the forces of supply and who as metallic money-owners simultaneously represent the forces of demand form a more integrated and cohesive, unitary merchant trading system which is more concretely unified in that each commodity is not just abstractly related to the general equivalent but is instead related to a specific quantity of gold. Moreover, just as in Hegel's *Logic*, quantity, in the form of money is now beginning to subordinate the previously established qualitative commercial relations. Whereas, in Hegel, determinate being was revived in quantitative form, here, the trading community is reintegrated as a more concretely unified commercial system through the operation of the society-wide market to which commodity-owners now relate more explicitly as merchants.

The use-value of money is not a product of its sensuous materiality but results from its abstract social virtue of general exchangeability. The setting aside of one particular commodity as "universal equivalent" or money generates an objective society-wide standard of value that can equate all commodities to each other in an integrated system. Now all commodities are bought with money and sold for money, and all values can be compared through money.

Money as means of purchase makes the exchange of commodities $C - M - C'$ possible, and, through this, establishes a market where exchanges become interconnected in such a way that the sale of any one commodity involves its purchase by another. At this point the social world of value connections is constrained, however, by the use-value wants of the exchangers who sell commodities they do not want to consume to get those they do. When, however, they get what they want, the exchanges stop. Thus, the motion of value is constrained by the consumption of use-values in the $C - M - C'$ circuit.

However, in money, unlike in ordinary commodities, the material contingency of use-value is held in abeyance. Although active money, which mediates commodity exchanges, cannot ignore use-values, idle money, which functions as a store of value, is not bound by this use-value limitation. When idle money is employed for the purchase of commodities, not for their mere consumption, but for gainful resale, the first form of capital emerges: capital as a self-augmenting motion of value.

With the formula $M - C - M'$ which buys a commodity in order to sell it again, money is withheld from the market, not to save up for some article of consumption needed by the saver, but to buy commodities in order to resell them for a profit. With this development, value can expand itself without the use-value constraints of $C - M - C'$ and thus transform itself into the primitive form of capital. Of course, this form of exchange will not take place at all unless the second M' is larger than the first M. Merchant capital takes on and off the forms of commodity (quality) and money (quantity) in a unified, unbroken operation of arbitrage. Just as quality and quantity in "the process of measure . . . pass over into each other" (*Logic*, 1975, 111) in an indifferent manner in Hegel's logic, so too does the underlying, operation of capital remain indifferent to the kind of commodity circulated, so long as its chrematistic principle is maintained.

A purely capitalist society may be said to be only externally represented by the circulation-forms of commodity, money, and capital which can exist independently of the capitalist production-process. These are forms attending commodity trade in general which, as Marx pointed out, always arises between self-sufficient economic communities. Commercial activity always arises through the intermediation of trade in which a merchant makes his profit through arbitrage or "buying cheap and selling dear". This activity is, however, both historically and logically prior to the capitalist mode of production. Commodity trade intermediated by money as well as capital, as a form of using idle money with which to "buy cheap and sell dear", existed in many societies prior to the emergence of the capitalist mode of production, and was operated independently of the capitalist production-process in these societies.

As in Hegel's realm of being in pure thought, the trading system is a self-determining system only implicitly or formally, since its material substratum lies outside its own motion. In the realm of the *forms* of circulation, no mention is made of labour as the *substance* of value. Yet the mode of operation of the trading system provides the basis for the reification of the labour-and-production process which forms the material foundation of all societies.

Capital would still be subject to the restrictions of use-values if its operation always consisted of the "buying cheap and sell dear" of specific commodities. Only when capital can produce *any* commodity that fits its chrematistic purpose does it become truly indifferent to all use-values. But to acquire this principle of self-determination, in the Hegelian sense, the commodity world must demonstrate its capacity to produce any commodity (quality) whatsoever, and hence achieve full "indifference" as to what use-value it is required to produce. To accomplish this objective, merchant capital must necessarily be superseded by *industrial capital* which has the capacity to subsume the labour-and-production process (essence) which is common to all societies under its sway, through the conversion of labour-power into a commodity.

In order for the autonomous forms of circulation to "subsume" the production-process they must also be able to subject that most essential ingredient of production, labour, to capital's own rules of behaviour. In other words, only when labour-power which is the source of productive labour becomes a commodity, and adopts the mode of behaviour peculiar to this commodity-economic form can the labour (and production) process be integrated into the chrematistic system of capital. Labour-power purchased as a commodity now becomes the source of the augmentation of value. This is the point of transition from the Doctrine of Circulation to the Doctrine of Production in the Uno/Sekine dialectic. Only in the context of production can the substance of value be explicated. Its real and logical premises in the Doctrine of Circulation have necessarily been presupposed.

7.4 THE DOCTRINES OF PRODUCTION AND ESSENCE

The development of the argument which makes possible the transition from the Doctrine of Circulation to the Doctrine of Production in the *Dialectic of Capital* parallels the transition from the Doctrine of Being to the Doctrine of Essence in Hegel's *Logic*, in which the dialectic of "quality, quantity and measure" ends with an "absolute indifference" to quality or determinateness (Sekine, 1984, 206–9). In each case, a diversity in "forms of being" are demonstrated to be the "outward show" or appearance of an underlying essence which must appear. Moreover, in each, the second doctrine examines the inner constitution of the subject matter and shows the way in which the inner content of the subject/object is subsumed by, and is reflected in, its existential mode.

Specifically, the Doctrine of Production demonstrates that the forms of circulation must arise from the production process of capital as the "forms of appearance" which the productions of capital take on. For example, commodities and money can no longer be external givens, by means of which capital performs its chrematistic operation, but products of the production process of capital itself. In other words, they are mediated forms and, thus, cannot be introduced from outside the motion of capital itself.

The first three sections of Hegel's Doctrine of Essence is entitled: "Essence as Reflection Within Itself". Essence is at first viewed as an unchanging substratum of being such that the latter represents what is inessential or a mere "seeming", but, when essence becomes existence, or, in other words, when essence is reflected in being, the underlying unity of the essence is split into two parts which face each other as polar opposites. Being is thus grounded in essence in such a way that essence has itself become self-contradictory. In the corresponding part of the Uno/Sekine economic theory, the "production-process of capital", a parallel development takes place when the form of industrial capital achieves "absolute indifference" because it is capable of producing any use-value as a commodity (i.e., as value). In this case the recurring contradiction between value and use-value takes the form of a contradiction between capitalism's value augmentation and the use-value production which prevails universally in any human society.

In the exposition of the labour-and-production process the Hegelian triad of "identity, difference, and the ground" is repeated three times, and each time the second term "difference" progresses through the stages of "distinction, duality, and opposition". First, the production of use-values in general is taken to be an identity, in which, however, the labour-process and the production-process can be distinguished as the two different elements. The unity of these two constitutes the labour-and-production process.

In order to establish the essence of capitalism, the Doctrine of Production necessarily introduces productive labour as the simplest concept, which, "despite its validity in all epochs", becomes abstract "not only as a category but also in reality" solely in the context of capitalist society.[8] Productive labour is at first, viewed as an identity. This identity turns out to be dual, consisting of a concrete-useful and an abstract-human aspect. From the concept of dual productive labour issues the notion of the expenditure of labour or labour-time, which is again considered at first as an identity. This identity, however, is bipartite, being composed of necessary labour and surplus labour. The concept of bipartite productive labour implies an opposition such that one part of it cannot be increased without decreasing the other.

The labour-and-production process common to all societies is, however, a socio-technical organisation without a self-driving force; an historically particular principle of management must operate this inactive organisation. "The ground is not yet determined by objective principles of its own" (Hegel, 1975, 179). From the point of view of the labour-and-production process common to all forms of economic organization, capital is not essential. Capitalism requires that production should be carried out under the chrematistic form of capital: The existence of capitalism means, that the labour-and-production process common to all societies or the ground is activated and unified by the specifically commodity-economic principle of capital (the law of value). Thus activated, or "grounded" the labour-and-production process becomes the process not merely of use-value production but of value formation and augmentation. Capitalists are thus no longer mere merchants but industrialists (i.e. executors of production, or, more precisely, purchasers and users of labour-power).

The working of the law of value or the production of value and surplus value, therefore, crucially depends on the conversion of labour-power into a commodity; it is this that translates the ground into existence. From the point of view of existence, labour-power as a commodity produces different surplus products in different capitalist operations. The development of the capitalist method of production, however, simplifies the labour-process and renders the mobility of labour virtually costless. Labour-power is also standardised and reduced to an indifferent, unskilled working capacity. The wage form as a category of cost now develops because the capitalist whose production no longer depends on specific but only on standardised labour-power readily purchasable in the market does not distinguish labour power from the other elements of production. With the commoditization of labour-power and the subsumption of the labour-and-production process to the process of value augmentation, capital becomes fully grounded and can thus contain the source of value augmentation within itself.

This conversion of labour-power into a commodity, it must be made clear, does not follow from the logic of transition or passing over from one form to another. Only the historical fact of this conversion, recalled in the past and internalized (*erinnert*), introduces the essence of capitalist production.

The consistency of commodity-economic principles with the general norms of economic life is demonstrated by a logic which follows or mirrors the Hegelian logic of "reflexion". The logic of reflexion or dependency shows how the object of study can contain the ground or foundation of its existence within itself so as to be self-determined. The $M - C - M'$ of capital as a circulation-form is not self-determined because it is not possible for everyone in a society to buy cheap and sell dear. It is only in the Doctrine of Production where $M - C - M'$ is transformed into $M - C(Lp, Mp) \ldots . P \ldots . C' - M'$ that it is made explicit that the commodities purchased (C) consist of labour-power and means of production suitably combined to produce the commodity for sale (C′) profitably.

The Hegelian triad of "the ground, appearance, and actuality" is translated into the following propositions of the dialectic of capital: (1) Capitalist production secures itself on the basis of the workers-versus-capitalist production-relation; (2) Industrial capital must circulate without interruption, while avoiding all unnecessary waste of resources; and (3) Capitalist society reproduces itself on an expanding scale, supplying means of production and articles of consumption in an appropriate proportion, while alternating between the widening and the deepening phases of capital accumulation. In so doing capital simultaneously reproduces commodities, and capitalist production relations. The essence, or inner constitution, of capitalism is thus laid bare, when the compatibility of use-value production in general and the specifically capitalist production of value (including surplus value) is demonstrated in (i) the production-process, (ii) the circulation process, and (iii) the reproduction-process of capital.

The dialectic of capital first shows that capital, the commodity-economic form of value augmentation indifferent to use-values, can successfully contain, at least in principle, the production of any use-value which capitalist society requires. The production of commodities as value is not yet ready to deal with the constraints presented by specific types of use-value in circulation. It must first demonstrate its ability to ensure the provision of use-values in general, since this constitutes the material foundation of economic life in all societies. Therefore, in the doctrine of production, capitalists are viewed primarily as purchasers and users of the single homogeneous commodity: labour-power. They are not yet distinguished according to the use-values by which they specialize to produce with differing industrial techniques. At this stage in the development of the theory, the organic composition of capital must be assumed to be the same for all capitalists because, only in this abstract context, can the law of value determine the real social cost of every commodity.

Once use-values are capitalistically produced as value, they must be circulated as commodities. Only by being embedded in the circulation-process of capital, is the production of commodities guaranteed of both its continuity and efficiency. This requires capital to bear circulation-costs of all sorts which, though necessary for the non-interruption of capitalist production, do not form or preserve value. These costs must be minimized if resources are to be devoted, as much as possible, to the production of surplus value directly.

The circulation-process of one capital presupposes that of other capitals, such that the motions of all separate units of capital make up an integrated motion of the aggregate social capital which systematically produces and circulates all commodities. It is this interconnected whole which constitutes the reproduction-process of capital; hence, this totality must be studied as a socially aggregate process.

The actuality of the aggregate social capital as a Hegelian absolute is undoubtedly the reproduction-process of capital or the capitalist *mode* of production as it reproduces itself. For capitalism to be actual it must first be shown to be, of course, possible. If capitalism is actual because it is shown to be one of the many possibilities, its presence is said to be "formally necessary" or contingent". If capitalism is shown to be actual because it satisfies all conditions that make it itself or in principle a possibility, capitalism is said to be a "real possibility" or "relative necessity". If, however, the presence of capitalism is shown to be self-determined, so that its actuality and possibility are no longer separable and if capitalism not only depends on its conditions but also produces these conditions of existence for itself, then capitalism is said to be an "absolute necessity" or "unconditioned actuality".

The formal stage of actuality may be interpreted to mean the reproduction of capital by capital itself contingent upon the conversion of labour-power into a commodity. For so long as, and to the extent that, labour-power is available as a commodity, the reproduction of variable capital is ensured; through this fundamental reproduction, constant capital and surplus value can also be reproduced, and consequently the capitalist production-relation as well. In the real stage of actuality, it is shown that the reproduction of goods by capital in the form of commodities satisfies all the conditions for capitalism to actualise itself. In other words, capitalist society must, as must any other viable society, produce all goods that are required for its existence. The reproduction-schemes ascertain the possibility of the capitalist mode of production. When the necessary becomes self-independent or self-conditioned, it achieves unconditioned actuality in the Hegelian dialectic. The actual process of capital accumulation is unconditioned in this sense. Capital no longer depends on a contingency; after it develops the law of relative surplus population. Then capital by itself makes labour-power available as a commodity. The law of value is thus no longer subject to any external restriction.

The compatibility of the reproduction of use-values and capitalist production-relations establishes the self-dependence of the capitalist mode of production in much the same way as Hegel's Doctrine of Essence, which, after proceeding through the dialectic of intro-reflexion, appearance, and actuality, ends with the category of "unconditioned self-dependence" or "absolute actuality", thus guaranteeing the self-sufficiency of the Absolute. When this has been established, we can proceed to the third doctrine, The Doctrine of Distribution, which parallels Hegel's Doctrine of the Notion.

7.5 THE DOCTRINES OF DISTRIBUTION AND NOTION

In both dialectics, the third doctrine exhibits what the object of study is by itself capable of developing once the consistency of its mode of existence and its substantive content is guaranteed. In the third doctrine of each dialectic, the logic of "development" (or self-fulfillment) lets the object of study unfold its working mechanism in an ideal environment so that it may reveal itself completely. Thus, in the Uno/Sekine dialectic, which has for its object of study a purely capitalist society, the Doctrine of Distribution shows how the capitalist mode of production develops its own market so as to produce all the diverse use-values which are socially demanded *as value* in a manner consistent with the self-adopted aim of capital.

In Hegel's dialectic, the Notion cannot be introduced until the Absolute establishes itself as actuality and requires no alien principles to unfold itself. The Doctrine of the Notion can then describe the subject-matter in its transparence (i.e. as "how it is because of what is"), following the logic of development or self-exposition. Since the "inner" and the "outer" of the subject-matter are already known, it only remains to show their reconciliation. The Notion, in its most developed form as the Idea, is the concrete or active universal which, unlike the abstract universals of the empirical sciences, generates and determines the particulars, which appear to be prior to it, from out of itself (through "practice of the will"). In other words, the self-differentiation of the concrete universal or Absolute Idea enables it to make explicit what is implicit within it, just as the developed plant is latent in its single cell (Dunne, 1977, 6). All the categories of Being (something, other, quantity, number, essence) and Essence (ground, appearance, actuality, reciprocity) are now seen as the self-differentiation of the Idea, and not as autonomous thought forms. Similarly, in the Doctrine of Distribution, the unity of the "outer" of being (circulation) and the "inner" of essence (production) shows that capital maintains its unity, through the establishment of a general rate of profit.

Up to this point, the dialectic of capital has delineated the concept of capital in its abstract operation, prior to its differentiation into particular spheres of production. Whereas, in the Doctrine of Production, value formation and augmentation are constrained only by use-values in general, and not by particular types of use-values, in the Doctrine of Distribution, capital is at last ready to differentiate itself into its heterogeneous forms to produce different use-values, while still preserving the unity of circulation and production established in the first two doctrines. Specifically, this involves the unfolding of the capitalist market. When industrial capital develops its own market in order to distribute surplus value, it produces commodities as use-values, while continuing to abide by its original principle of value augmentation. In thus reconciling the two sides of the production of commodities, as value and as use-values, the rate of profit acts as the subjective notion of capital. For the production of commodities as value by capital is already secured as objective fact by the time capital applies itself to the distribution of surplus value in a manner that is most congenial to its own rationality. In the present doctrine of distribution, use-values are no longer alien elements to industrial capital; they have become its necessary ingredients by the time it develops its own market. The dialectic of development that governs the formation of the capitalist market echoes the Hegelian triad of "universality, particularity, and individuality". The general principle of the distribution of surplus value must be preserved as it adapts, or conforms, to the manifoldness of use-value production. Thus industrial capital is "universal" as it produces all use-values, "particular" as it produces different use-values, and "individual" as it forms a unified whole producing different use-values.

Capital is not only differentiated into branches of industry with distinct organic compositions dictated by differing use-value considerations which in turn dictate different production techniques, but, as well, into commercial and interest-being capital. Hence, in this context, the resolution of the contradiction between value and use-value involves reconciling the diversity of the use-values which are required by the market and which necessarily involve different methods of production with the indifference of capital to the production of commodities as value. This differentiation is unified, however, through the law of average profit, by which capital ensures that surplus value is distributed in proportion to the magnitude of money value of capital advanced, while the actuality or self-dependence of capital is guaranteed by its law of population.

The Hegelian triad of the subjective notion, the objective notion and the Idea can now be translated into the following propositions of the dialectic of capital:

1) Specialized units of industrial capital, producing different use-values, determine equilibrium prices in the capitalist market, so as to divide surplus value among themselves in the form of average profits.[9]

2) Part of surplus value must be ceded as rent to private landed property not only to guarantee the principle of equal opportunity to all units of capital but also to ensure capital's access to land in general.

3) In order to save unproductive costs of circulation, capitalist society develops mechanisms of utilizing idle funds in the most efficient manner for value augmentation, but the consequent division of average profit into interest and entrepreneurial profit externalizes the relations of capital so that even capital itself can, in principle, be converted into a commodity.

It should be emphasized that the economic forms of profit, rent, and interest cannot be properly understood, and demystified, without a prior knowledge of the Doctrines of Circulation and Production, which explain the necessary inner connections of capitalism. In the absence of the knowledge provided by these two doctrines, the rate of profit appears to be a mere mercantile form indifferent to the productive base of society, since the market automatically effaces any trace of a specific mode of production and establishes a universal relation of equality among traders.

Without a prior understanding of these doctrines, no explanation can be given as to why rent makes it appear that a thing, land, produces value; while interest, the most fetishized form of all, makes it appear as if capital can automatically create value independently of the production-process. The Doctrine of Distribution demonstrates that these fetishized forms are simply the outward manifestation of a self-contained and self-determined inner logic. In the Doctrine of Distribution, this inner logic of capital reveals itself, not in the sense that it tends to become transparent, but in the sense that the theory has been developed to the point that it now is possible to make the connections between its outward manifestations and its inner logic.

Hegel's Doctrine of Objectivity is in close correspondence with the theory of rent in the dialectic of capital. The process in which "the Notion determines itself into objectivity" is, according to Hegel, "identical in character with the ontological proof of the existence of God". Hegel's observation that, "God, as absolute spirit, is known only in his activity; in his works",[10] may be translated into the language of the dialectic of capital as follows: if the distribution of surplus value in the capitalist market is comprehended only from the point of view of capital (as in the theory of profit), it remains the Subjective Notion and is neither "realized" nor "objectified". The distribution of surplus value as profit cannot, therefore, always remain a strictly internal affair of industrial capital. The production of commodities as use-values cannot occur without land, nor can capital ignore land's private ownership which ensures the separation of the direct producers from their natural means of production. Hence, it is necessary to understand the distribution principle in its activity or operation as it involves a factor alien to capital, namely, landed property. Capital must adapt its method of distributing surplus value, while ensuring that its law of average profit is preserved, so as to accommodate the participation of landed property in the capitalist market.

In the conversion of surplus profit into differential rent of form I, landed property merely represents the differential fertility of lands to which capitalist activity is largely extraneous. Capital considers landed property as an external entity to which it must transfer permanent and indigestible surplus profits as rent. This function of landed property in capitalist society is not unlike the Hegelian "mechanism"; it remains wholly abstract and mechanical. The second relation that capital establishes with landed property echoes the Hegelian relation, "chemism". Landed property now presupposes capitalist action, and then takes advantage of it. Now landed property not only compels capital to produce maximum surplus profits, but also deters long-term investment. The third and last relation that capital concludes with landed property echoes Hegel's "teleology". Landed property no longer represents nature against capital, but justifies itself as a necessary ingredient of capitalist society. This is immediately apparent when landed property directly participates in the pricing of the capitalist market, so as to collect absolute and sometimes even monopoly rent, but the cunning of capital is such that it successfully constrains the greed of landed property by assuring its increasing affluence as capitalism develops itself.

Since the production of use-values involves land, the natural means of production monopolized by landed property; capital, in order to acquire a right to the use of land, cedes part of its surplus value to landed property in the form of rent. The concession of rent to landowners enables capital to develop a new self-image, the self-image of interest-bearing wealth. In thus recognizing the forms of rent, capital, in effect, accepts the idea that a non-owner of value-objects can also, in some cases, have a rightful share of surplus value. Although this alien idea, is initially imposed on capital, capital willingly applies it to itself to free itself from all use-value restrictions.

Loan-capital and commercial capital also partake of surplus value by saving industrial capital its circulation-costs. They thus earn a share in surplus value, much like landed property, by contributing, if only indirectly, to the production of surplus value. When this distributional method is generalised in such a way that capital earns interest in much the same way as land earns rent, there arises the purely reified form of interest-bearing capital, which sublimates the production of value and surplus value by industrial capital. Just as Hegel's dialectic of "subjectivity, objectivity, and the idea" was able to develop the notion of the absolute idea as "the idea which thinks itself", the notion of capital, through the form of interest-bearing wealth, is able to return to its origin in the sphere of circulation as the "absolute idea of capital".

In the sphere of circulation, the form of merchant capital was only a primitive form of the Notion, for it was not a self-mediating process; in the doctrine of production, capital became a self-valorizing process, and, hence, a more adequate expression of the notion of capital. But it is only in the Doctrine of Distribution that capital concretizes itself in a manner which allows particular capitals, through competition, to establish a general rate of profit which governs all spheres of the commodity economy: industry, agriculture, commerce and banking. The dialectic ends when the notion of capital, which governs all forms of circulation and production, has progressed through the categories of profit, rent, and interest until it itself becomes a commodity wholly relieved of any material content, the use-value of which is a pure self-augmentation. Capital which automatically bears interest cannot stop circulating, for to stop circulation would be to forego the interest that can be earned.

With this result, the system of the dialectic of capital becomes a self-enclosed whole. All previous determinations of capitalism are sublated in this result, which simultaneously returns us to the beginning of our investigation of capitalism. The commodity, in a fashion not unlike the Absolute Idea, has taken us full circle by displaying all its antecedent categories as the self-differentiation of its eternal operation. Thus, the "commodity" is the simplest logical category or abstract-universal, which anticipates the genesis of capital, but it is also the most synthetic logical category or concrete-universal in which capital finds its ultimate expression.

A purely capitalist society which constitutues the inner logic of capital is now completely exposed, without leaving a thing-in-itself unknown. It reveals itself as a self-generating, self-concluding and self-revealing totality, with a necessary beginning, a necessary unfolding, and a necessary closure. The dialectic of capital is thus the self-explaining logic of capital, as both object and subject. The dialectic of capital reveals the inner logic of capital in toto; that is to say, it exposes the whole of what is (capitalistically) rational and objective in capitalism.

7.6 CONCLUSIONS

The dialectic of capital is possible because capital has its own logic which the economist can copy directly. Yet this logic also reproduces many features of Hegel's logic which coincides with metaphysics.

Sekine has demonstrated that there is an almost perfect correspondence between the dialectic of capital and Hegel's *Logic*. In this case, logic coincides with economic theory. A dialectic of capital is possible because the theorist can allow the purification and reification of capital to complete itself in theory, and thereby to arrive at concepts as purified of contingency as Hegel's philosophical universals. Hegel and Uno use the same dialectical method of total comprehension except that the subject/objects are different. If Hegel's subject/object was "Reason" (or the divine wisdom), the subject/object of the Uno/Sekine dialectic is "capital".

It is commonplace among Marxists that Marx rejected Hegel's idealism, but it is equally true that Marx re-read Hegel's *Logic* while preparing *Capital* and found it very useful. for his political economic study of capitalism. The following passage suggests just what it was that Marx rejected in Hegel and why he, nevertheless, found Hegel's dialectic useful for the understanding of capitalism. As Marx explains,

> . . . I do not start from "concepts" and hence do not start from the "concept" of value, and . . . I do not have to "divide" the latter in any way. What I start from is the simplest social-form in which the labour product is represented in contemporary society, and this is the "commodity". I analyze this . . . first in the form in which it appears. Here I find that . . . it is in its natural form, a thing of use . . . a use value, on the other hand, a bearer of exchange value. . .. Further analysis of the latter shows me that exchange value is only a "phenomenal" form, an independent mode of representation of the value contained in the commodity, and then I proceed to analyze the latter (*Capital*, 1971a, 108).

Marx is critical of Hegel because Hegel "transforms the process of thinking into . . . an independent subject . . . the demiurgos of the real world, and the real world is only the external, phenomenal form of the idea" (Marx, 1971a, 19). Furthermore Marx believes,

> Hegel fell into the illusion of conceiving the real as the product of thought concentrating itself . . . and unfolding itself out of itself, by itself, whereas the method of rising from the abstract to the concrete is only the way in which thought appropriates the concrete [in reality] reproduces it as the concrete in the mind. But this is by no means the process by which the concrete [or real] itself comes into being (1973, 101).

As a good materialist Marx recognizes that the commodity economy as a material and social reality does not come into existence merely to allow value, as the absolute Idea of capital to unfold its logic or reveal its inner essence. To believe that it did would be to succumb to an idealist illusion similar to that which misled Hegel. While in the case of the specific reality of capitalism Marx cannot accept the illusion that "the real is the product of thought" he does recognize that to understand or appropriate the concrete reality of capitalism, an economic system which reifies social relations as commodity relations, it is necessary to reproduce in thought "the categories [and laws or logic] which makes up the inner structure of [that] society". In other words, he recognizes the necessity of a Hegelian or dialectical approach to understanding capitalism which entails "evolving the different forms" of bourgeois society "though their inner genesis" (Marx, 1973, 101, 108) to arrive at a synthetic or concrete knowledge of that object.

Uno and Sekine have carried on the work that Marx left unfinished when he died. They have developed the dialectic of capital as a rigorous materialist science of capitalism. In fact, their development of the dialectic is not only more rigorous than Marx's, but is, in fact, more consistently dialectical than Hegel's. They are able to accomplish this feat because the concept of "use-value" which opposes "value" in their materialist dialectic has a substance which is lacking in Hegel's "nothing" which he opposes to "being".

Following Sekine, a dialectic is meaningful only when the overcoming of the antithesis by the thesis is not a foregone conclusion from the beginning. Hegel's *Science of Logic* starts with "being", which represents the presence of the Absolute, and "nothing", which represents the absence of the Absolute. But, if the material world is created by the Absolute in accord with Reason, then what can the non-being or absence of the Absolute mean? By contrast, in the dialectic of capital, value, the formative element or being of capitalism, stands opposed to use-values which represent the material factors of economic life devoid of commodity-economic specifications. Whereas value implies the presence of the commodity-economic organizing principle in society, use-values represent the real economic life of society which must be carried on before, during and after capitalism.

With the eventual abolition of value (capitalism), use-values (human societies) remain, whereas with the abolition of Hegel's Absolute, there can be nothing (none of His creations) left. Thus, "being" is never really constrained by "nothing" in Hegel, and the triumph of the former over the latter is predetermined before analysis. In other words, the absence of the Absolute does not appear to have any other meaning than total void. But, if that is true, the overcoming of "nothing" by "being" lacks any real substance. A dialectic which progresses by overcoming recurring contradictions posed by real use-value resistance is not just a dialectic of categories but a scientific and materialist one based on the self-purifying or reifying tendencies of capitalism, and the use-value resistance to these tendencies which genuinely exist in historical reality.

While it is true that in the natural sciences, the principle of non-contradiction is fundamental, the dialectic of capital presents an objective basis for a social science based on dialectical contradiction. Because capitalism is an inverted and reified reality, it has the capability to largely remake the world in its own image. Capitalism carries within itself a tendency towards the realization of a purely capitalist society, making it possible to both envision such a society in thought and to theorize its necessary inner connections as a dialectic.

The objectivity of the dialectic of capital stems from the logical "self-containedness" of capitalism as an economic system and not merely from the brute fact that some vaguely specified social institution existed historically for some finite period. The dialectic is no less real because it is an organizing principle rather than a directly observable and empirically visible phenomenon. Because this logic must always contend with varying degrees of human and use-value resistance in any historical society, capitalism's laws operate only as tendencies. A stage theory of capitalism's historical development must, therefore, be theorized to mediate between the pure theory and empirical studies of historical capitalism. A stage theory must explain which events in the history of capitalism's growth, maturity and decline were the necessary results of the operation of the logic of capital and which events were the consequence of contingently appearing human and use-value resistance to that logic. Kozo Uno and other members of the Uno School have developed such a stage theory but a discussion of that topic lies beyond the scope of this article.

A final cautionary note. I have been attempting to show a correspondence between the Marx/Uno logic of capital and Hegel's logic of the Absolute. The scientificity, objectivity and validity of the dialectic of capital is, however, in no way derived from or dependent on its correspondence with the Hegelian schema. It is derived from the theory's own ability to reproduce the logic which tended increasingly to govern the economic life of English society in the middle of the 19th century, when the commodity-economy was permitted to regulate a greater and greater portion of the material economic life of society.

Notes and References

1. For an interesting discussion of the relation between the laws of capitalism which are revealed in the dialectical theory of pure capitalism and the general norms of economic life, I highly recommend Sekine's article, "Economic Theory and Capitalism", *York Studies in Political Economy*, I (1982).
2. Sekine's *Dialectic of Capital* (1984), begins with a long introduction. In the latter part of that introduction (60–86) Sekine discusses the relationship between capitalism in theory and capitalism in history. He also defends the necessity of both a pure theory and stage theory of capitalism and gives a devastating critique of the logical-historical method.
3. In the introduction to his *Dialectic of Capital*, referred to in the previous footnote, Sekine is careful to distinguish between the dialectical approach to capitalism and theories of capitalism which are ideal types (*Dialectic*, I, 18–49).
4. The expressions such as the subject matter, object of study, subject-object, theoretical object are used in this paper interchangeably in the same sense.
5. My account of the basic features of the Hegelian and Uno/Sekine dialectics was greatly influenced by Sekine's account in his introduction to the *Dialectic of Capital*, (1984, 26–59), Robert Albritton's discussion in Chapter 7 of his book, *A Japanese Reconstruction of Marxist Theory* (1986), and, finally, by Patrick Dunne's discussion in his unpublished manuscript, "Hegel's Doctrine of Quality in reference to the Theory of the Commodity-Form in the Dialectic of Capital".
6. Uno, Sekine and Albritton discuss the sequence of categories in the pure theory of capitalism and the sequence of categories in history (Albritton, 1986, Part I; Sekine, 1983, pp. 70–77 and throughout the Uno text, Uno, 1980). A good short introduction to this and related topics occurs in article by Sekine in the *Journal of Economic Literature*, "Uno-Riron: A Japanese Contribution to Marxian Political Economy" (1975), pp. 847–77.
7. With regard to the parallels between the Doctrine of Circulation and the Doctrine of Being, I rely heavily on Dunne (1977) whereas in my accounts of the correspondence between the Doctrines of production and Essence and the Doctrines of Distribution and the Notion I rely on Sekine, *Dialectic of Capital*, vol. 1 (1984) and vol. 2 (1986). I should point out that for purposes of establishing a correspondence with the dialectic of Hegel, I refer to Wallace's translation of the *Logic* (1975).
8. Dunne, (1977) p. 6; Sekine, 1984, pp. 257, 264–7; Hegel's *Logic*, pp. 389, 479.
9. Commodities are exchanged among capitalists at prices diverging from values, due to the fact that the social and private costs of an individual commodity are not the same whether in money terms or in terms of labour. The divergence of prices from values is an expression of the inner logic of the capitalist market which arranges social production by means of private chrematistics. Values cannot be discarded or ignored merely because we can now speak of prices as well. Value continues to relate the commodity or market exchanges which take place in an historically

transient capitalist society to the material economic production of use-values which must go on in all societies, whether capitalist or not. If there were no distinction between the formal operation of the market and the substantive aspect of economic life, the question of relating them would not arise. Only then would value be redundant or unnecessary as is claimed by some followers of Sraffa.

10. Sekine, 1986, pp. 285–8; Sekine, 1984, p. 55.

8 Postmodern Sediments and the Logic of Capital

Randall Terada

> "Consider that a sedimentation of gender norms produces the peculiar phenomenon of a 'natural sex' or a 'real woman' or any number of prevalent and compelling social fictions, and that this is a sedimentation that over time has produced a set of corporeal styles which, in reified form, appear as the natural configuration of bodies into sexes existing in a binary relation to one another."—Judith Butler, *Gender Trouble.*

The critical robustness of recent theoretical claims that call for the recognition of the breakup of the stable authoritative "I" and for genealogical inquiries into its production in discourse, have made groundbreaking inroads in feminist theory and politics in general. In Marxist circles the influence has been more . . . . hesitant, perhaps owing to the claim that once this theoretical novelty is allowed into the hallways of Marxist theory, one is impelled into a "post-Marxist" terrain. But a recognition of the internal history of Marxist debates reveals that the Frankfurt school in particular placed working class subjectivity on the agenda a long time before the French post-structuralists issued their decentring decrees. Socialist feminists in turn were working at the myriad of problems of an historical materialist dialectic that overlooked the realm of household production and gender identities. So to be fair the recent postmodernist critiques were nothing new to some Marxists already increasingly self-reflexive about the applicability of its conceptual apparatus.

Nevertheless central to what could be called a "postmodern" turn in theory is a growing awareness that these 'supplements': women's labour, race, environment, sexuality, served to consolidate and uphold crucial binary oppositions that structured critical paradigms like Marxism. The claim was made that struggling to build a community built upon communal class processes is a necessary but not sufficient condition to thwarting sexist and racist exclusionary social practices. For Marxists the deep structure of the capitalist juggernaut reveals an economic logic that is decisive and terrifying in its singularity. It underpins and complexly "determines" the superstructure or surface circulation of cultural codes which on the surface reveal a high-tech carnivalesque disorganization complete with a cast of clown politicians and cyborg bankers.[1] On this surface all cognitive mapping is permanently skewed, "all that is solid melts into air". All is ersatz and flux. All that is, except for the enlightened "I" which conspicuously escapes all interrogation in this otherwise free fall of sushi, satellites and simulacra. But can this Marxist topography hold any longer?

The core of what has been labelled an 'essentialist' Marxism has been the preponderant weight attached to class subjects whose 'interests' can be determined from the positions they occupy in the relations of production. This construction assumes the homogeneity of this subject and effaces its raced and gendered identity. Recent interventions have taken essentialist theory to task for typically structuring the social totality into an ontological dualism.[2] This ontological dualism can take many forms: between the ephemerality of surface appearances and a deeper underlying rational structure; between an inner realm of necessary laws of motion and an outside of contingency and historical accident; between some form of multicausality which stipulates contingent, conjunctural factors of causation on one hand and an ultimately determining set of core factors or invariable elements on the other.

Thus the dualism articulates a conception of the social totality in which there are deep essences which are the providence of theoretical analysis to grasp. These essences structure reality, either directly or indirectly, but a theory's "correctness" depends on its methodological appropriation of the invariable essences that structure the social whole. Explanation is then reduced to that of effecting an adequate theoretical synthesis between two incommensurate ontological spheres: the deep structures of necessity and the contingent flux of the surface.[3]

The combination of these two aspects: a subjectivity placed within a closed and centred social totality, leads to formulations that infer an inherent privileging of certain forms of subjectivity, causal factors and necessary inner laws. The result is that social relations become intelligible only by referring them to their underlying essence which provides them with an ultimate fixity (Bertramsen, 1991, 8). In Marxist theory this "ultimate fixity" constitutes a founding centre which structures the social totality while itself escaping the process of structuration. When the "economy" serves as this founding centre one has arrived at the *sine qua non* of an essentialist Marxism. Any critique of essentializing tendencies in Marxist theory must begin by questioning the inherent presupposition of the unity of the economy. The economy can not be seen as an *a priori* unified space that subsequently acts on other regions of the social totality. This essay will argue for the more productive anti-essentialist analysis which resists the depiction of the economy as "a homogeneous realm governed by endogenous logics" but make the counter-claim that it is rather a "heterogeneous terrain, cracked and fissured by political struggles" (Daly 1991, 18). What all too often happens is that politics is too often subordinated to a logic of capital, with the consequence that political struggles and interventions that cannot be explained in terms of classes, economic laws or fundamental structures are seen as historical contingencies that "interrupt rather than modify the unfolding logics of capital. As a result they are either discarded, or treated as untheorized supplements to the fundamental laws of historical development" (Bertramsen 1991, 17).

What is needed is a way in which to combine the analysis of micro-struggles that are local and contextual with a knowledge of the macrostructures of oppression. A combination which includes a micro-analysis sensitive to the local, but in doing so, does not expel from the picture the larger structurating forces. A theory, for example, that can provide the "logic of capital" as one such macrostructure without belittling "difference", and that maintains an anti-foundationalist epistemology.

Recent Marxist attempts at multicausality such as the important and substantial work of Ernest Mandel are prone to essentialism simply because they do not displace the ontological dualism they are ultimately trying to escape. As an example of how an ontological dualism structures a discourse of Marxist crisis theory, we shall briefly consider Mandel's important work on long waves of capitalist development.[4]

In his opening chapter of his work, *Late Capitalism* Mandel clearly outlines his theoretical stance which purports to emphasize the multiple causes of "long-term tendencies of development and the inevitable collapse of the capitalist mode of production" (Mandel 1978, 34). Mandel's emphasis here on multiple causation places him in a long line of Marxists who have attempted to introduce complexity and to avoid reductionism in their analysis of capitalism.

Mandel states that the necessary laws of capital can explain the transition from economic expansion to stagnation, in other words forecast capital's crisis tendencies. But most importantly, once an expansionary long wave is initiated then the laws of capitalist development, which Mandel has heretofore held backstage, are brought up front and locked into place. On the flipside only the basic laws of motion of capitalism trigger each wave's depressive phase; the transition from long wave expansion to long wave contraction is a necessary one which follows from the "internal logic of capitalist laws of motion".

His emphasis on "multiple causation" is delimited solely to the period from stagnation to expansion. Mandel argues that the initiation of a new long wave expansion above all, "cannot be deduced from the laws of motion of the capitalist mode of production by themselves" (Mandel 1980, 21). The upturn is contingent and results from "non-economic factors" (gold discoveries, major innovations, or capitalist victories over the working class) which activate the counteracting tendencies to the falling rate of profit.

By maintaining that each economic long wave is sustained by the inner laws of capitalism, he drastically reduces the theoretical space allowed for the intervention of human agency and contingency. Instead he insists that

> . . . to deny that once a new long wave is under way the inner logic of capitalism (i.e. the laws of motion of the system) must of necessity command the further trend of events is to deny that these laws of motion are operative in any real sense whatsoever (1980, 21).

The meaning of this passage seeks to reaffirm in a classical Marxist fashion the privileging of the economy and establishing its *a priori* causal effectivity in the hierarchy of the social totality. By privileging the logic of capital in this manner Mandel serves to discipline the heterodoxy that issues from any other discourse that seeks to undermine the homogenous space on which the economy resides and uncover what could be termed the process of its political constitution. But if this disciplinary move is not enough Mandel goes on to excoriate any "non-economic" determinants that would purportedly render the economic logic illogical.

> If one believes that not just once every fifty or sixty years, but continuously, external non-economic forces determine the development of the capitalist economy, then one rejects out of hand Marx's entire economic analysis (Mandel 1980, 29).

In sum Mandel constructs a theory that allows for the consideration of exogenous factors but ultimately reduces these factors to only those that contribute to counteract the tendency of the rate of profit to fall. Once these factors combine to allow a expansionary phase of the long wave to begin they are spirited aside and Mandel relies on an abstract conception of capital's inner laws of motion. In the end historical development is explained as a function of the inner laws of capital (Norton, 1988).

The working of the necessity-contingency dualism in Mandel is quite straight-forward. For Mandel history is a combination of the forces of necessity and contingency. If necessity drives the system into crisis, it then loses its sway over the events of this crisis, and contingency is then brought front stage centre. But there is only a limited theoretical space for contingency to operate. Only when the necessitous laws of capitalism have propelled the system into crisis, does Mandel mark this as the moment for the intervention of human agency and contingent social forces.

The split between the forces of necessity and contingency serve to grant Mandel's theory a neat and tidy separation between an inviolate Marxist science, that is, the study of the laws of motion of capital, on the one hand, and the contained space of human agency, the contingent forces of history, on the other. Only when the inner laws fall into contradiction and run aground into crisis does the opportunity arise for human intervention. Mandel's notion of agency is secondary to the workings of the inner laws of capital. Once capital is placed on track again, it takes on a will of its own, outside of human intervention, until it comes crashing down again in approximately the next fifty year period.

The ontological dualism here is evident in that there exist two well defined frontiers, the problem for theory is to specify the jurisdictional limits of each and to maintain the internal consistency of the prior region (necessity) from the destabilizing potential of the latter (contingency). In this manner the primacy of the first term, the logic of necessity, is secured from any systemic threat from an "outside". The only way in which the "outside" affects the integral logic of the realm of necessity is when this logic runs into its own self-contradictions. In other words, to maintain the possibility of a Marxian science of the economy, Mandel restricts agency, contingency, struggle, to a secondary role, whose intervention is merely supplementary to the more overarching logic of capital as it runs its course through history.

The usual correction to this problem is to privilege the second term of the necessary/contingent dualism; that is, to read into the logic of necessity the very contingency that was expelled in order for the logic to function. Thus what happens is that now all possibility for structural articulation and necessity becomes conditional upon agency. Privileging the second term invites only a type of voluntarist solution. Class struggle, for example, is read right into the logic of necessity, deforming at every turn its will to objectivity.[5] Thus the popular Marxist debates then focus around the dynamic of this ontological cleavage with opposing camps dismissing each other's claims on the grounds that they insufficiently pay heed either to agency or to structure. Additionally debates revolve around which elements to include on which side of the binary. By leaving intact the necessity-contingency dualism, these debates revolve around the extension/retraction of the two frontiers, and their respective contents (DeMartino, 1993). This whole debate peaked in E. P. Thompson's attack on Althusser's efforts to place *Capital* on a more scientific footing by purging it of its humanist elements, provoking Thompson to resuscitate the agent, consciousness and will in history and reinvoking the familiar humanist privileging of the centrality of conscious human agency. The debates then remain within the theoretical terrain of the binarism and attempts to break from this have been few.

8.1 OVERDETERMINATION AS A SOLUTION?

This section will briefly look at the attempts of Resnick and Wolff to pose a different solution, that is, not to reverse, but to displace the dualism altogether. Their initial work has proven quite influential, becoming the basis of a working group, Association for Economic and Social Analysis (AESA).[6]

Resnick and Wolff's attempt to escape the binarism which structures the discourse of essentialism is based on an epistemology of overdetermination. An overdeterminist epistemology contests the above two claims of the necessity/contingency dualism in the following manner. Overdetermination makes no claims of essential causes, in fact it argues that reality is composed of a processual dynamic, which is plural, and always transgressing and blending into mutual realms of the social such that one can only speak of relative processes of the cultural, political and economic, but not of determinate fixed regions. The blurring of all boundaries of the social results from a conception of the overdetermined nature of the social totality such that every cause is at the same time an effect and vice versa.

> Overdetermination transforms the idea of causality. It becomes futile to try to explain the cause of A by searching for the essential B and/or C and/or D that best or most explain it . . . Overdetermination begins instead with the presumption that event A is caused by innumerable influences emanating from all the other events in the social totality. In principle . . . the full or final explanation of A's causes is impossible (Resnick and Wolff, 1988: 53).

The complex combination of a number of overdetermined processes that congregate and construct all objects of knowledge leads to the conclusion that theory can never provide a "true adequate, comprehensive, and complete account of their essential causes". Thus any depth analysis of deeper structural causes of a phenomenon that structure and determines "surface appearances" is suspect if one accepts the humbled status of theory in this overdetermined account of social reality.

In sum their proposed solution to essentialism is to refute the ontological dualism it is premised on by disbanding the hard and fast frontiers and rejecting the attempts to classify elements of theory based on this distinction between necessity and contingency. The epistemology of overdetermination rejects essences, hierarchies, determinate causes and effects, and in its place it conceives a strategy that focuses on continual flux, and partial explanations that do not seek grand "truths" but are merely the "stories" about how one particular process, for instance the class process, may contribute to an explanation at a particular local site, overdetermined by a number of other processes. The effect is that social analysis becomes not whether one may approach closer the absolute nature of the "truth" of society, but having given up such "fetishes", focuses rather on how this particular overdetermined narrative, i.e., of class processes in the household for example, can make linkages with other radical narratives that focus on the overdetermined nature of the social (Fraad, Resnick and Wolff, 1989). In this way, Resnick and Wolff escape the charge of having produced a metanarrative of the social since their particular entry point into the social, that of class, is meant to contribute to other democratic struggles and radical narratives and thus affecting and being affected by this overdetermined relation with these discourses. Nothing remains unaffected by this complex inter-relationship. Both theoretical knowledges and practical struggles emerge transformed. Thus having renounced any absolute truth claims dependent upon knowledge of an essence of the real, Resnick and Wolff claim that Marxists can contribute a particular knowledge to the anti-capitalist struggles, knowledge about particular class processes at innumerable sites in society.

This theoretical outlook has produced some extremely fruitful radical analyses of contemporary politics. They emphasize class as a process of production, appropriation and distribution of surplus labour, rather than as an objective identity of persons or group of people with determinant interests occupying a particular position in the relations of production. Class is seen as a process in which agents participate at a number of sites in society. For example they identify the household as one site of the social in which distinctive class processes of a feudal type take place. Most importantly by identifying class processes as not a determinant cause, but a partial and overdetermined effect of a number of other processes: gender, economic and political processes, they succeed in providing a sophisticated anti-essentialist analysis of oppression in the household structures in the contemporary United States. In their analysis they undertake to include struggles over the meaning of gender identity, conservative political agendas that prevent women from receiving equal pay as their male counterparts, laws and informal practices blocking women's access to birth control and abortion, cultural processes that reassert stable meanings around notions of woman's relations to their own bodies and self-images, and traditional notions of gender as regards marriage, household tasks, and parent-child upbringing.

> Gender processes affirming women's inferiority do not necessarily or automatically relegate women to the household and to housework. The latter must themselves comprise a socially devalued sphere for the woman, as gender devalued, to be assigned to them. Other cultural processes must rank household production and childcare as less important, less prestigious, and less productive. Then the conditions are in place for the feudal fundamental class process to combine with the inferiority status attributed to women to consign them to the role of feudal surplus labor performers (Fraad, Resnick and Wolff, 1989, 27).

Here it is clear the care they take to mention other processes which together create the dense network of material relations, meanings and agencies that coalesce at a particular site, the household in this case, and place women in an exploitative relationship.

Notwithstanding the positive benefits this theoretical approach contributes to Marxian theory, there are two problems with the epistemology of overdetermination. The first has to do with the possibility of explanation. By displacing the necessity-contingency dualism in the name of overdetermination, what happens to concrete analysis premised on a examination of causal processes? The second deals with the charge that this school has not escaped the very essentialism which it claims cripples the explanatory power of other theories.

Firstly: in order to combat the essentialist theory that separates the social into two spheres of necessity and contingency, the overdetermination school breaks with this problematic by proclaiming the impossibility of locating any determinate cause and effect. They abolish causal analysis outright, which begs the age-old question about the baby and the bath water. Must one, in order to abolish deterministic analysis, blur the relations between elements to such an extent that all one can do is compose partial analyses of an object that deny any possibility of claiming that some processes may be more determinant of an object's existence rather than another. They state that "The task of Marxian theory is to construct a knowledge of an ever-changing overdetermined social totality" (Resnick and Wolff, 1987, 97). But this task the AESA school sets out for itself has proven to be extremely difficult. Most of the attempts made by this school that have tried to "explain" an event must violate the very strictures of overdetermination, because in order to study something even partially, the overdetermined, ever-changing complex of processes must be in a sense "frozen", and overdetermination itself must be absented "so that some sort of causal relation may be constructed" and studied by the analyst (DeMartino 1992, 29). To arbitrarily claim half way through an analysis that the reason that something is this way and not another, is because it is "overdetermined" is to deny the theoretical project itself, which is to provide explanations, not necessarily determinate causes and effects, but reasons why one sees it this way and not another and be able to reasonably defend that position. In other words there must be an attempt at synchronic analysis if one is not to be swept away in a diachronic wave of processes upon processes.

Secondly, as DeMartino deftly points out, the crux of the problem with overdetermination is that it sets about an impossible task because it itself has not escaped the anti-essentialist strictures in which it judges the fidelity of other theories (DeMartino 1992, 32). Resnick and Wolff are vehemently opposed to any type of essentialism "in" or "of" theory

(Resnick and Wolff, 1987, 96). They clearly label both empiricism with its assumption of a singular reality that can, through trenchant observation, be grasped in theory, and rationalism, which reduces the concept to the essence of the real, as two essentialist epistemologies. But their insistence that overdetermination is an extra-discursive attribute of objects which theory then must map, thus violates the basic tenets of their research program which would eschew any such direct access to the "real".

Thus Resnick and Wolff have not themselves escaped an epistemological essentialism. In their attempts at analysis they treat overdetermination as an objective property of extra-discursive objects which they then attempt to "reflect" or capture in anti-essentialist social theory. Beginning with the proposition that overdetermination is an attribute of extra-discursive objects that anti-essentialist theory must reflect if the latter is to be deemed adequate to its object thus violates their strict code that theory is only a partial analysis of an always overdetermined process.

In their work, then, overdetermination, as an objective property of social reality that must be copied in theory, parallels the very structures of the above two epistemologies which Resnick and Wolff dismiss as essentialist. Resnick and Wolff fall into the same essentialist trap that they adeptly identify and critique in other theories.

8.2 DISCOURSE THEORY AS A CORRECTIVE TO THE ESSENTIALIST DUALISM

The discourse of essentialism is premised on an ontological dualism of the type that Derrideans ravish like vultures to a fresh kill. Within this binary opposition the pole of necessity is an attempt to fix the social as a stable system of differences. The task of theory then, is to point out the extra-discursive realm of objects which are stable and which lie beneath the surface but which nevertheless establish a determinate pattern, a logical necessitous pattern to social life, as Therborn writes:

> . . . science is the systematic study of a specific object: that this object has the character of a pattern of determination or of regularities; that the emergence of a science of society, such as economics, involves the discovery of a specific social pattern of determination, the functioning of which it is then the task of science to analyze (Therborn 1976, 101).

Theory when used properly is meant to engage in an analysis which captures this level of objects and the logical and necessary relations between these objects thus determining the effectivity of the "real".[7]

The attempt by theorists like Resnick and Wolff and Laclau and Mouffe to contest this essentialist discourse borrows from a postmodern refrain which argues for the "discursive nature of all objectivity". Laclau and Mouffe attempt to show that, in the history of Marxism, the concept of "hegemony" has expanded in the discourse of various Marxist thinkers in order to articulate the growing breach that occurs between the logical and the historical or necessity and contingent. Faced with the abundance and contingency of history, they argue that reading the inner motion of capital directly into the struggles and antagonisms of daily life rendered all efforts of analysis incredibly problematic. Arguing that identities cannot be reduced to positions in the relations of production and that laws of motion of capital cannot be reduced to concrete history, they trace the concept of hegemony as an attempt to bridge the growing divide between a logical unfolding of an immanent necessity and the contingency of politics.

For Laclau and Mouffe then, society, conceived as a centred and closed totality reducible to an underlying rational logic forecloses on the radical possibilities that present themselves in a different guise other than those forms that are strictly regulated by an economistic logic. Under the banner claiming that "society is impossible" they argue that it can only be constructed and continually be reconstructed out of the political strategies of different forces in society, each striving to undermine differing programs while attempting to implement their own "logic of the social". Society for them, is ultimately a contingent construct:

> If society had an ultimate objectivity, then social practices, even the most innovative ones, would be essentially repetitive: they would only be the . . . reiteration of something that was there from the beginning . . . But if contingency penetrates all identity and consequently limits all objectivity, in that case there is no objectivity that may constitute an "origin" . . . "Articulation" in that sense, is the primary ontological level of the constitution of the real (Laclau, 1990, 184).

Objectivity, rather than something "discovered" by the theorist is constructed via attempts by various social forces to arrest and re-articulate the interminable instability of the social around a system of stable signifiers or nodal points.

The biggest difference between Resnick and Wolff and Laclau and Mouffe is that where Laclau and Mouffe (1985) argue from the initial premise that discourse is the attempt to arrest the contingency of the social around a stable hegemonic pattern of meaning and signification, Resnick and Wolff do not claim any stability or potential stability at all, since as we have seen, all objects exist in a state of overdetermination. As argued, what happens is that for Resnick and Wolff the ability for a social analysis is drastically limited since they emphasize the processual nature of the social, and thereby downplay stable patterns and sedimented social practices amenable to social analysis.

Similarly for Laclau and Mouffe, their attempt to escape the ontological dualism that structures essentialist discourse is to rely exclusively on the discursive construction of the social, and the "impossibility" of the social to construct itself as a stable system open to objective study. Rather than pose a determinate level of stable objects and logical relations, they argue for the ultimate contingency of every level of the social or as Laclau notes, "The movement towards deeper strata does not reveal higher forms of objectivity but a gradually more radical contingency" (Laclau, 1990, 182–3). The difference between Laclau and Mouffe and the AESA school, is that while the former acknowledges that the social can be stabilized around discursive 'nodal points', the latter reject any such possibility that objects of theory can be stabilized in a hegemonically consolidated pattern.

What cost do Laclau and Mouffe incur by this exorcism of all traces of essentialist discourse from their analysis? Reading their dense prose one might claim intelligibility. But more substantially, Douglas Kellner, Linda Nicholson and Nancy Fraser and a host of others, have recently cued upon the inability of discourse analysis to recognize the existence of macro-structures of oppression and most importantly, for our purposes, the macro-structure of capital.

> Discourse analysis tends to ignore the fact that the social clearly consists of different layers with varying degrees of stability (or sedimentation). This means recognizing that to assert that everything is, in principle, open to change does not establish whether or not actual social relations and institutionalized features of capitalism are all equally changeable to the same extent (Bertramsen, 1991, 163).

What this important point reveals is that anti-essentialist analysis runs the opposite risk of its essentialist opponents, that is, its emphasis on the discursivity of all social objectivity, reduces everything to a flatness where all is equally transmutable and open to counter-discursive practices.

What discourse theorists disregard is the plausible assumption that the social may in fact be a "layered" object, perhaps ultimately contingently grounded, but nevertheless bounded by different levels of necessity. But does recognizing a differential necessity to the social, force us to concede to an essentialist analysis? Perhaps a solution to this problem should seek not to reverse but to displace the ontological dualism of essentialist discourse, but at the same time avoid the problems that accompany the claims made above by AESA and Laclau and Mouffe.

Thus far we have outlined the problems specifically with the postmodern moment in Marxist theory: (a) its inability to fix within its discursive field a stability and causal effectivity amenable to social analysis and (b) even when it attends to the principle of a fixity of the social based upon the principle of 'nodal points' which are principle signifiers that manage to hegemonize the discursive field around a set of articulating principles as noted in the theory of Laclau and Mouffe, one runs into the problem that a *purely discursive approach cannot deem to grasp the abiding structural regularities that undergird a fixity on the discursivity of the social.* In other words discourse analysis cannot recognize that the social is layered into relatively more or less stable patterns of solidity. The sale of labour power for example is much more "sedimented" as a social practice than say, voting behaviour. To put it differently, worker struggles around change are usually focused around the amount of surplus value the capitalist allocates as part of the wage basket, rather than over the very sale of that labour power to the capitalists. The difference here is huge. That struggles in Western industrial countries have focused around the former rather than the latter is testimony to a more deep seated embedded structural necessity that favours a logic of capital notwithstanding the periodic struggles over wages. The argument here being that discourse theory does not recognize the difference between the two since to do so would ultimately focus on a level that does not accord with the emphasis on discursive "surfaces".

This leads us into recognizing the relative "sedimentation" of the social into various levels or the overriding historical embeddedness of certain structural forces. It is the Uno School that has made some headway in theorizing this layering of the social. It provides an outline of an analysis of the logic of capital, and how it conditions the existence and participates in the overdetermination of other objects constructed at other levels of analysis.

8.3 THE UNO SCHOOL

No doubt the postmodern fascination with the "linguistic turn" has raised the ire of many political economists, Marxists and mainstream alike, who envision, with this shift, the eclipse of an underlying rationality in theorizations of the social. This view insists that the postmodern move towards flux and its focus on language and texts is somehow neglecting the "real".[8]

Uno's prescient insights anticipated much of the recent trend towards the semiotic theoretical bend in left social sciences.[9] He accurately pointed out that *Capital* as it stands is a text that is steeped in reductionist tendencies and thus promoted the notion of a "levels of analysis" approach. What this in fact resembles is a concerted effort to "deconstruct" the very text that Marxists had up to that time considered a sacred text, inviolate in its very historical significance and revolutionary veracity.

An Unoist "deconstruction" of Marx's classic text was deemed necessary on account of what we have heretofore been investigating, that is the ontological dualism at the heart of essentialist analyses. Uno perceived that Marx had inadvertently mixed levels of analysis in his approach to the study of the capitalist mode of production. In his study of the logic of capital, there is included a mish mash of conjunctural politics, political history, ideological excoriations with their accompanying "truth" claims, working class struggles, economic theory and a theory of the constitution of the working class subject.[10] All these different discourses interfered with what Uno considered to be the dialectical logic forming the basis of the theory of capital.

He thus went about "taking apart" the veritable castle that Marx had constructed and extracted the rational kernel of the text or what his student Thomas Sekine later termed the "dialectic of capital". But this construction of the dialectic was not possible prior to renting the fabric upon which Marx laid down his concepts, and fundamentally re-ordering them. In other words he re-ordered the relational logic of Marx's work so that, for instance there would be no confusions over barter as opposed to commodity exchange or that the substance of labour was not introduced before the commodity logic was firmly in place. What Uno recognized was that the signifying chain that Marx placed down was in need of a different relational logic in order for it to make "sense", that is, in order to overcome certain reductionist tendencies in his text.[11] In so doing Uno's thought paralleled to a certain extent that of the linguist Saussure in that in any text there are no positive terms only differences, identity is wholly a function of differences within a system. Thus in regards to his relationship to *Capital*, Uno's intervention disrupted the traditional logical sequence of categories which had run into contradictions and could no longer be resolved, and placed them in a new differential sequence. He deconstructed a veritable shibboleth, and continued on to reveal that not only had his method gutted and laid to rest certain fundamental reductionist problems of Marx's text, but also there stood the possibility, something post-structuralists deny, for a creative re-construction of Marx's work. In other words after his deconstructive efforts he forged on, past the ruins and incoherent fragments of logic, history, politics, ideology that lay at his feet, and rebuilt to a large extent a version of Marxist theory that displaces reductionist and essentialist analyses via a levels of analysis approach.

The Uno school ascribes to Marx's three volumes of *Capital*, a specific knowledge of a particular macrostructure of power, or what is called in particular, the logic of capital. The logic of capital is a distinct dialectic which is a totally closed and centred system, based on a logic of strict necessity and fuelled by the central contradiction between value and use-value. This system is closed in that it is restricted to only the elements which enable the logic to construct a "commodity economic logic". This logic constructs the absolute identity of capital in its total transparency. Essence directly coincides with appearance. There is no room for the elements so constructed at this abstract level to be articulated differently, no chance of "free floating signification" but rather evoking an absolute narrative, the story of capital as told by capital.

This may sound much like a sophisticated reformulation of the necessity-contingency dualism. The Uno school attempts to escape this with their subsequent adoption of a levels of analysis approach. The level of strict logical necessity is not a copy theory of an extra-discursive referent. It is a radical constructivist theory of capital. Capital is not a correspondence to anything "real" outside of itself. Although at this abstract level one can posit its historical conditions of existence, what cannot be read into this level is any explicit extra-discursive referent. This is signalled by the Uno School commitment to levels of analysis which entails that the logic of capital is no mere reflection of an extra-discursive object, i.e., history or the economy.

> . . . a single abstract logic may be countered by all sorts of other forces that may deflect it, and in particular cases nullify its effectivity all together at the level of the concrete. In short, one logic does not empirical reality make (Albritton, 1993a, 31).

In other words, there are counteracting tendencies, forces that constitute an 'outside' to the pure internalism of the dialectic which constantly work to deform it, preventing it at the level of theory from finally ever reaching a stable objectivity that expands to all levels of the social.[12]

However instead of reacting by deeming everything processual and overdetermined, or relegating everything to a discursivity, the Uno school promotes three relatively autonomous levels of analysis which draws on the fact that "reality" is not a continuum that can be gradually apprehended by one consistent epistemology, but is discontinuous.[13]

This way of constructing Marxian social theory has many merits. It seems to be able to better grasp at once the macrostructural logic of capital, yet at the same time is neither reductionist nor essentialist. One may recall how Resnick and Wolff, ironically fell into this trap of attempting to reflect the overdetermined nature of reality directly in theory. Uno theory recognizes the ontologically discontinuous nature of reality, is committed to anti-essentialism on all fronts, in theory and of theory, and draws attention at lower levels of analysis to the discursive and overdetermined nature of all objects. In addition it escapes the type of radical relationalism that befalls the AESA school and Laclau and Mouffe by recognizing that the logic of capital is sedimented into different forms at the level of history. The Uno approach has made headway by its insistence on the macro-structural effects of one precise object, that is capital.

> The purpose of Marxian economic theory (the dialectic of capital) is to "lay bare" and comprehend the whole logic of capital. It does not observe and evaluate the operation of capitalism "out there", from the outside, and formulate various hypotheses which may be susceptible of empirical testing . . . *the logic of capital cannot be learned by the alternating steps of tentative theorising and its empirical confirmation*, as in natural scientific studies (Sekine, 1993: 8, my emphasis).

The knowledge of capitalism is not an algorithmic process. Its knowledge comes about through a process of theoretical production. To this end the Uno school, not unlike Althusser, proposes a new way in which to read *Capital.* To read this work is to recognize that the precise object of study is constructed discursively and is not a simple reflection of something "out there". The three volumes of *Capital* are not, in other words a theory of the genesis of the capitalist mode of production, nor is it a narrative of class struggle, economic crisis or British society for that matter. To read *Capital* the Uno School tells us, is to break with all notions of expressive realism that posit a unproblematic relation between theory and a direct and simple relation with an extra-theoretical world of objects. Once this is recognized, we then intervene and combat the metaphors and tropes of liberal economics with an arsenal of our own and re-articulate the relational discursive logic that combine liberal political doctrine and free market economic theory, with a narrative that replaces the benign order of capital, with the reifying effects of the "logic of capital".

8.4 CONCLUSION

Having outlined in a very general way the broad contours of Uno theory, by way of conclusion it should be noted how this theory attempts to overcome the ontological dualism inherent in essentialist theory. Firstly it seeks to overcome this dualism by (a) displacing the assumptions that provide this dualism with its cohesiveness and meaning and (b) provides one with the space to view both macro and micro analysis without necessarily sacrificing one for the other.

For the sake of a final comparison, the assumptions governing the necessity/contingency dualism will be briefly reviewed here. Recalling that the dualism rests upon a foundation that separates the social into two distinct realms: the former providing stable objects, inner motions and laws, while the latter is all that is expelled from this prior region of necessity. Both the AESA school and Laclau and Mouffe contest this assumption by forwarding a theory of overdetermination and discursivity of the social respectively. Uno theory does not differ in kind, but only in degree from certain aspects of these theoretical projects. Uno theory shares the break with all forms of rationalism and empiricism. It shares the emphasis in both AESA and Laclau and Mouffe on the discursive nature of knowledge and the "real". Yet one can also discern where there is located in Uno theory a determinate break from this turn to the postmodern: (a) in its location of a macro-structure of power in the logic of capital and (b) its more sophisticated notion of the layered nature of the social.

Overdetermination as a condition of existence of objects is not applied to all objects in Uno theory, on the contrary it recognizes that one object, specifically capital, is susceptible to a closed analysis whereupon its overdetermined relation is held in abeyance until its specific logical nature is exposed for critical analysis. This critical analysis reveals the macro-structure of capital, a narrative which highlights the tendencies of capital in a form free from outside contingencies subsuming everything to a commodity-economic logic. Most importantly instead of assuming that the social is divided between two ontologically incommensurate realms, separated by a "frontier which mutually limits their range of effects" (DeMartino 1992, 86), Uno theory blurs the distinctions radically. The sphere of necessity, unlike Mandel who represents this as an objective feature of reality, is in Uno theory a dialectical logic that does not refer to an extra-discursive referent. The objects constructed at this level are articulated with the *sedimented* forms of the social at a lower level of analysis. The levels of analysis approach, unlike the more discursive and overdetermined analyses mentioned above, recognizes the different levels of the social such that it cannot be grasped in one overarching methodology. The middle range of analysis, *stage theory*, constructs its object, a particular stage of capital accumulation, via a more structural logic that explores regions of sedimentation of the social. Bertramsen explores the nature of this concept of sedimentation:

> Over time the social becomes increasingly *sedimented* as its political "origin" is gradually forgotten . . . social agents encounter a number of recurrent patterns of meaning and action through their engagement in day-to-day activities . . . If left undisturbed these recursive practices will result in an increasing institutionalization and routinization of the social. Hence, we shall say that the social has become sedimented. However, the political "origin" of the social is *re-activated* in so far as its sedimented forms are contested, for example, in the case of dissent, conflict, civil disobedience (Bertramsen, 1991, 30).

In the Uno approach this level of analysis, *stage theory*, accompanied with an abstract knowledge of the logic of capital, constructs the way in which a particular configuration of social, economic and political and cultural forces cohere in a regulatory pattern, i.e., are sedimented, thus anchoring a particular hegemonic project that supports and provides the conditions of existence for, among other things, an accumulation strategy for capital. Having studied these sedimented effects, historical analysis, the third level of analysis, constructs the genesis of the various regions of sedimentation of the social as outlined at the level of stage theory. In other words it locates their "political" origins.

From the above it has been argued that the Uno approach has made some modest theoretical headway in a variety of ways. However one note of caution must be exercised as to the nature of its theoretical project. One way in which the levels of analysis approach is attempting to overturn the essentialist epistemological project is its focus on the construction of its objects *within theory*, that is, the logic of capital produces an abstract inner law of motion; stage theory produces a structural outline of the way in which an accumulation project is sedimented via various struggles in the political, cultural, political and economic sites of society over a particular time span; historical analysis traces the genesis of this particular accumulation strategy and uncovers the political origins of this sedimented political-hegemonic project. It is a more fluid analysis of change and micro-political struggles.

For example one can approach a particular object, say theorizing the post-consumerist construction of neo-liberal subjects as reflected in the ideology of popular media and the discourse of political parties and think tanks, using a levels of analysis approach. Here the object is not as susceptible to three methodologically distinct levels of analysis in the same way as say, the study of the changing workplace in Southern Ontario auto manufacturing plants in the twenty year period between 1973 and 1993. The former study, to avoid reductionist analysis, i.e. seeing the interpellation of subject positions as solely fodder for corporate capital, would use stage theory and the logic of capital solely to establish a very broad contextual framework for the analysis. The latter study of manufacturing worksites, would benefit from a different, perhaps more "synergistic" relationship between the three levels of analysis. In sum, the relationship between the levels cannot be determined *algorithmically* (with *a priori* methodological protocols) but one must instead rely on the *verisimilitude* of the relationship between the levels of analysis and the object being constructed.

The levels of analysis approach can be said to be founded on the anti-essentialist moment in Marxist theory. To this end it makes more complex the relationship between theory and its objects. This paper has attempted to come to grips with such a legacy in the Uno approach through a comparison with other efforts to overcome the dualism in essentialist theory. Seeking to problematize the nature of the Marxian theoretical enterprise with the claim that there are no objects beyond theory in some pure sense means that the nature of the 'real' is nothing short of discontinuous. To make this claim is to admit of a postmodern moment in Marxist theory. To claim anything short of this may very well be to languish on the rocks of orthodoxy and to rehearse worn out claims that can say nothing more about the future course of the Marxian theoretical project(s)

Notes and References

1. For example according to David Harvey once one acquires a sound understanding of the motion of capital and its structural requirements, the flux and seeming ephemerality of the social becomes understandable not as something *sui generis*, but as effects of a logic of capital accumulation which has entered a new phase:

   > Even though present conditions are very different in many respects, it is not hard to see how the invariant elements and relations that Marx defined as fundamental to any capitalist mode of production still shine through, and in many instances with an even greater luminosity than before, all the surface froth and evanescence so characteristic of flexible accumulation (Harvey 1989, 188).

   Flexibility, flux, decentred identities are merely referred to, and disciplined by, an organized capitalism on a higher order: the power of capital on a global scale (117, 159). For an interesting reply see Gibson-Graham (1993).
2. The term "ontological dualism" that structures essentialist discourse was brought to my attention by a paper presented by George DeMartino at a *Rethinking Marxism* conference in the fall of 1992.
3. DeMartino explains this boundary dispute as follows:

   > The boundary separating these incommensurable realms is arbitrary and unstable. The initial tracing of the boundary, which is understood . . . to be necessary for economic science, must actually *precede* the science itself (1992, chap 6).

4. For this section on Mandel I am much indebted to the argument put forth by Bruce Norton (1988).
5. For an elaboration of this argument see Albritton (1986).
6. In this section the work of AESA and Resnick and Wolff will, notwithstanding differences amongst some of them, represent one general position, as such their names will be used interchangeably.
7. Besides Goran Therborn, *Science Class and Society* (1976), David Harvey *The Condition of Postmodernity* (1989) is also an exemplar of this method.

8. The turn to a post-modern/structural analysis has done its best work in feminist theory as opposed to the more arcane things done in the American academic English departments. Feminists in various areas have made headway in developing some of the more radical implications of this theoretical focus, see for example, Butler (1990), Cornell (1991), and Fraser and Nicholson (1990).
9. Postmodern theory no doubt has its weaknesses especially in the French nihilist bent. Baudrillard (1981) and Kroker (1991) exemplify the imbalance in the minds of some postmodernist thinkers who seek to revel only in contingent and arcane "meaninglessness". Similarly we have witnessed how Resnick and Wolff and Laclau and Mouffe respectively have succumbed to "discursive overload". But this is by no means reason to reject certain insights provided by postmodern theory. In fact some aspects of postmodern theory can strengthen Uno theory while Uno theory can contain some of its excesses.
10. For an interesting analysis of a Marxian theory of the constitution of the subject see Amariglio and Callari (1989).
11. Amariglio and Callari echo a similar interpretation of Marx's text:

    > All discourses (not only Marx's, but also Smith's or Ricardo's or the discourse of neoclassical economists) can be read as comprising different orders in which signifiers are articulated into discourse in order to produce different meanings; that is, different discourses are different constitutions of signs rather than different interpretations of an empirically given object of analysis. . . . Thus for example, the meanings that Marx ascribed to his key concepts (e.g., value, economy, ideology) do not have fixed, empirical referents. These meanings are construed as they are because Marx, facing the meanings produced by classical political economy, confronted this school of thought – its politico-theoretical priorities – by setting out to produce a *different* set of priorities (Amariglio, Callari, 1989:42).

12. Objectivity in the standard sense refers to the possibility that social being can be fixed in a stable set of relations that reveal themselves to the analyst via proper methodological protocols. Such is not the definition of "objectivity" used in conjunction with Uno theory, especially as outlined by Sekine. To the extent the Uno school make claims for objectivity in theory, it is not along the science/ideology continuum, but instead is meant to refer to the transparency of the logic as exposed at the most abstract level of the dialectic of capital. In this sense it is an objectivity as revealed as total necessity, objectivity as transparency prior to its deformation at other levels of analysis. Thus to the extent they hold this deformation possible their definition of objectivity does not reside in the subject/object dualism. To put it another way, "objectivity" is a signifier that is fought over and rearticulated away from positivism and metaphysics, and placed with the relations of a different language game; that being for instance something like Rorty's notion of solidarity (Rorty, 1985).

13. Resnick and Wolff rely on a singular ontology of overdetermination, similarly Laclau and Mouffe argue that contingency and negativity are the ontological postulates of all social reality. For a discussion of the Uno position which differs from these see Albritton (1993a).

9 Theorizing the Realm of Consumption in Marxian Political Economy

Robert Albritton

In recent years Marxian political economy has been criticized from a variety of points of view for privileging the realm of production and for largely ignoring the realm of consumption or for seeing it more or less as a passive reflection of production.[1] While there is some truth in this criticism, the problematic nature of the onesidedness is to be found less in Marx's *Capital* than in the manner in which Marx's famous text has been interpreted and utilized. For many interpreters the abstract economic laws of Marx's *Capital* are to be applied directly and without mediation to concrete socio-historical reality.[2] As a result, if the realm of production is primary in abstract economic law, they assume it must always be primary in the same way and to the same extent at the level of empirical social reality. But this does not follow. If we conceive of the laws of motion of capital as a theory of pure capitalism and leave open the question of the extent of their impact and efficacity in particular concrete social settings, then it by no means follows that the relation between production and consumption at the level of the concrete will mirror the relation at the level of abstract theory. In so far as a concrete social setting is capitalist, we would expect production to be important, but at the level of the concrete, capitalism never has the purity that it does in abstract theory, and consequently at the concrete level the realm of consumption may relate to production in complex ways. What I shall argue in this paper is that if Marx's *Capital* is considered to be chiefly a theory of capital's inner logic or a theory of pure capitalism, then production is clearly determinant over consumption, which more or less passively adapts, at least at this abstract level of analysis.[3] And although the primacy of production in capital's innermost identity suggests that such a primacy is indeed characteristic of capitalism in its inner logic, at more concrete levels of analysis not only might the primacy of production be considerably attenuated, but also the conceptualization itself of these spheres and their articulation with each other may make categories like "primacy" inapplicable.

Part of my motivation for writing this paper is the fashionableness of thinkers like Baudrillard, who use the seeming arbitrariness of present-day clothing fashion as a kind of paradigm for conceptualizing commodities as a system of signs whose meaning is constituted solely from a system of difference a la Saussure.[4] Baudrillard (1975, 121) has interesting things to say about the cultural and symbolic dimensions of commodities and consumption, but the overall effect of his conceptualization is to collapse economics into culture or perhaps both economics and culture are flattened out into a system of signs governed by "The Totalitarian Code". As chains of signifiers, commodities for Baudrillard seem to have lost their rootedness in use-value.[5] I suppose if there were a contemporary Marie Antoinette influenced by Baudrillard, she might say: "let them eat signs".

In this paper I shall not carry out a systematic critique of Baudrillard, but a concern to challenge his ideas does constitute a kind of background to my theoretical efforts.[6] My approach to theorizing the realm of consumption is informed by Japanese Political Economists Uno and Sekine's idea of three levels of analysis.[7] According to Uno, Marxian Political Economy should be understood as the science of capitalism, and this science at the most abstract level is constituted by a theory of capital's inner logic, at the level of mid-range theory by abstract types of capital accumulation most characteristic of different stages of development, and at the most concrete level by the analysis of historical change from the point of view of capitalism.[8] The three levels are not related to each other through simple deduction or induction, and although exploring their complex interrelations in detail is beyond the scope of this paper, my discussion of the realm of consumption at each of three levels will give at least some indication of how I conceive of interlevel relations.[9]

9.1 CONSUMPTION AT THE LEVEL OF PURE CAPITALISM

A purely capitalist society is a society conceived from the point of view of a self-regulating commodity-economic logic, where all production is production of capitalistically produced commodities. It is not a concrete society in the sense of giving full scope to use-values and humans, and this is because social relations only enter into the theory in so far as they can be managed commodity-economically by self-expanding value. But social relations can be so managed to a maximum extent if all inputs and outputs of production are totally commodified, the crucial commodity being labour-power. Total commodification assumes that in the context of capitalist profit-making there are no direct person-to-person relations, but instead that all connections between persons are mediated by things: namely, commodities, money, and capital. Or to use Marx's short-hand metaphorical expression, social relations are reduced to a "cash nexus".

To the extent that the economy is totally commodified, individual subjects are ultimately reduced to being bearers of economic categories. Thus a particular capitalist, no matter what his or her concern for workers or the environment must make an adequate profit or go under. Or a particular worker, may quit working for a particular capitalist, but he or she must eventually work for another capitalist or starve, and competitive pressures will tend to make wages and working conditions similar amongst capitalists. This is because it is assumed that in a purely capitalist society all production is the capitalistic production of commodities with three and only three classes: capital, labour, and landlords. The state, the family, and ideology are in the background, but cannot be theoretically specified except in very formal terms at this level of abstraction because they may involve either direct person-to-person relations or the use of extra-economic force. Or, in other words, they are not fully managed by a commodity-economic logic.

In a purely capitalist society, sociality is constituted by the motion of value. Or to say the same thing in different words: social relations are totally reified (commodified). Value is that which all commodities have in common, namely, that they are the products of homogeneous, abstract human labour. Value is what makes them exchangeable, and exchange is the only form of social connection. As values commodities are only signs in the narrow economic sense in that through their mutual interaction they signal the division of the total social labour amongst the production of various types of commodities.

As use-values commodities are conceived of as material objects with qualitatively different properties. It is at this point that Baudrillard launches his critique, arguing that use-value is just as social as value because ultimately it depends on a system of needs that is socially constructed as a sign system.[10] In opposition, it is my contention that no extensive theory of needs can be developed at this level of abstraction. All we can say in the context of pure capitalism is that certain things by virtue of their material properties are socially wanted, but we cannot specify which things or why they are wanted except in very broad categories determined not by the needs of concrete individuals but by the needs of self-expanding value.

Labour-power is a commodity that is wanted because it has the use-value of being the universal source of value. Land is a commodity that is wanted because it has the use-value of being the basis of all production. Capital in the form of interest-bearing capital is a commodity that is wanted because it has the use-value of generating profit. In the reproduction schema, Marx distinguishes between those use-values that enter into the individual consumption of workers and capitalists and those that enter into the productive consumption of capitalists.[11] For the sake of producing profits capitalists consume the use-values of labour-power and means of production. In order to live workers consume a wage-basket of use-values, and in order to live well, capitalists consume a boat-load of use-values. But at this level of abstraction, we cannot specify whether or not beaver skins are wanted for coats, or beef is wanted for hamburgers, or gasoline is wanted to propel autos.

Historically and socially specific wants and needs are at least in part determined culturally, but the study of meaning and symbolism requires the consideration of direct person-to-person relations, something that is absent in pure capitalism. Thus the cultural dimension of use-value and consumption must be studied at a more concrete level of analysis. In so far as we want to theorize capital's inner logic, it is appropriate to consider use-value primarily as a material and qualitative aspect of commodities by virtue of which they are wanted. The sociality of use-value is basically empty in this context, except in so far as value gives it a content as in the cases mentioned above such as labour-power. In other words, we have no way of knowing which use-values are socially desired except for those broad categories which are directly determined by the value relation. Thus, for example, labour-power is wanted by capitalists because it creates value and surplus value. But we do not know whether or not workers purchase horses, bicycles, motorcycles, automobiles, or public transportation passes in order to get around. Furthermore, persons become personifications of economic categories, and hence can only be specified in broad categories such as "workers" or "capitalists". We cannot specify gender, age, race, ethnicity, religion, marital status, nationality or any other characteristics of persons. Indeed, all persons fall into three and only three social categories: capitalists, workers, or landlords. Landlords own all the land, capitalists own all the capital, and workers sell their labour-power. Needs and wants cannot be specified other than noting that a minimum of use-values must be consumed by all persons in order to survive. Beyond that we do not know anything about the typical range of use-values consumed by capitalists (either productively or individually), workers, and landlords.

Use-value, then, is not theorized in the law of value considered as capital's inner logic as a relation between a subject, a need, and an object.[12] Indeed, this is the way in which Böhm-Bawerk (1949) and marginal utility theorists typically conceive of use-value. The starting point of marginal utility theory is usually the isolated needy individual casting dollar ballots in a cafeteria of use-values. Use-value in Marx's law of value has almost nothing in common with such an approach. In the law of value, individual subjects are objectified by being subsumed to capital as self-expanding value. Needs cannot be specified except for those determined directly by the motion of value, and hence there is no role for an independent theory of needs in the law of value. The actual range of use-values produced cannot be specified except that they must divide in such a way that necessary means of consumption and means of production are available. Thus the claim made by Sahlins (1976, 170) that "by correlations in a symbolic system pants are produced for men and skirts for women" does not apply at this level of abstraction. In the context of a purely capitalist society we cannot specify whether men wear pants or skirts or nothing at all.

Given this interpretation of Marx's law of value, it is possible to show clearly why it is that the sphere of production must be considered primary and the sphere of consumption secondary. The overriding aim of capital is to maximize profit, and profit is created entirely in the realm of production. If we understand capital as essentially self-expanding value, then it is crucial to understand that value can only expand itself by subsuming the labour and production process. Marx expresses this primacy of production by claiming that capitalism involves "production for the sake of production".[13] This is an important claim because it stands in sharp contrast to the notion of "consumer sovereignty" so common to most economic theory informed by marginal utility theory. It is based on Marx's view that capital produces only with a view to profit, and only discovers after the fact whether or not what it has produced is wanted. Furthermore, if we examine the motion of capital through a crisis cycle, we can see plainly that consumption must passively adapt to the expansions or contractions in the sphere of production. In the phase of prosperity capital is expansive, reducing the industrial reserve army and pushing up wages as a result of increased demand for labour power. According to Marx, during this phase, workers can expand their consumption. Once a crisis sets in and production contracts, wages fall and so does consumption. Thus it is accurate to say that in a purely capitalist society the realm of production has clear and determinant primacy over the realm of consumption. Production is the active center of profit-making, while consumption passively adapts to it. Furthermore, the realm of consumption remains relatively unspecified because we do not know the range of use-values that enters it except for the broad categories determined by the requirements of self-expanding value.

9.2 CONSUMPTION AT THE LEVEL OF STAGE THEORY

Whereas at the level of pure capitalism, economic life is tightly unified by being subsumed to the necessitarian logic of self-expanding value, such is not the case at the level of stage theory. Because stage theory is developed at a more concrete level of analysis, we can no longer assume the total commodification of all inputs and outputs of capitalist production. At this level the economic has political and ideological conditions of existence, and is therefore theorized as a set of relatively autonomous practices. For example, historically labour-power has always resisted being commodified with the result that its commodification has never been total. Stage theory, then, must concern itself with the typical degree and extensiveness of the commodification of labour-power at different stages of capitalist development and how this degree and extensiveness was maintained through not only economic means, but also ideological and political.

Furthermore, at the level of stage theory we need to specify the typical range of use-values capitalistically produced in the core capitalist country or countries. For it is precisely the specification of the most typical capitalistically produced use-values that is the basis for theorizing a stage-specific mode of capital accumulation. Thus the putting-out production of woollens was most typical of the stage of mercantilism, the factory production of cotton by industrial capital was most typical of the stage of liberalism, the production of steel by finance capital was most typical of the stage of imperialism, and the production of autos by transnational capital is most characteristic of the stage of consumerism.[14]

At the level of capital's inner logic, production, circulation, distribution, and consumption were integrated into a tight logical totality with production being the dominant moment.

At the level of stage theory, this logical structure is translated into relatively autonomous institutions, where it is no longer so clear where the primacy lies. Not only is the economic determined in important ways by the political and ideological, but also the economic itself becomes differentiated into relatively autonomous spheres such that the primacy of the realm of production over the realms of circulation, distribution, and consumption is not so clear or determinant. Because of the centrality of profit-making to capitalism, the realm of production is bound to be important, but it is conceivable that in a stage of truly mass consumption that consumption itself may become integrally bound up with profit-making in the sense that very active political and ideological interventions may be required to continually expand consumption (effective demand) for the sake of profit-making.

How the realm of consumption is constituted and how it relates to the realm of production varies from stage to stage. Since my concern at the level of mid-range theory is principally with stages of *capitalist* development, my focus is still not on entire societies, but is primarily on the most typical forms of capital accumulation in the most capitalistically developed country or countries. Following is a sketch of some of the differences between the constitution of consumption in different stages of capitalist development.

In the stage of mercantilism, the most characteristic form of capital accumulation was the woollen putting-out system, and putting-out workers consumed little beyond food, clothing, and shelter.[15] Also much of what they consumed was either not capitalistically produced or was produced by production relations that were only embryonically capitalist. Since in the putting-out system the home is also the basic unit of production, the labour of production and the labour of consumption would not always be clear and distinct. There was not much domestic labour to do, and it is likely that husbands and children shared in it to some extent since there was a premium on female spinning labour that for technical reasons generally could not keep up with male weaving labour.

In the stage of liberalism, when the basic unit of production shifts from the home to the factory, domestic labour and the labour of consumption becomes clearly separated from the labour of production.[16] I will not try to trace out the implications of this, but clearly such a change is bound to have a great impact on the organization of consumption, which now becomes largely separated in time and space from production and which to a much greater extent is the consumption of capitalistically produced use-values.

In the stage of imperialism, which reached its most classical expression in Germany and the U.S. in the twenty years prior to World War I, we see the early signs of modern consumerism develop with some increase in real wages, some increase in the array of capitalistically produced commodities purchasable by workers beyond food, shelter, and clothing, and the first steps in the development of modern mass media and advertising. Capitalists attempted to stem the class struggle and worker militancy that developed in this stage through the development of "scientific" management and through the development of paternalistic schemes aimed at buying worker loyalty. The most characteristic form of use-value production shifts from cotton to steel. In the two most advanced areas of the steel industry, The Ruhr in Germany and what is today called the "rust belt" in the United States, large steel corporations typically provided low cost housing and built certain amenities for the community. It should also be pointed out, however, that health and safety conditions were horrendous in the U.S. with about 25% of the unskilled workers in the steel industry being injured each year (Brody, 1960, 101). Also workers in the steel industry were expected to work twelve hours a day, 365 days a year (assuming no lay offs which in fact were frequent). This work schedule would not have left a lot of time for shopping (it would have to be done by wives to a large extent) and other activities associated with consumption.

In the stage of consumerism, for the first time, industrial workers of the leading industrial nation, the United States, can afford to buy such a complex and expensive use-value as a car. Indeed, in this stage as opposed to all previous, the realm of consumption comes to assume an importance that rivals the realm of production. Furthermore, both politics and ideology intervene actively to ensure that the realm of consumption continues to expand as much as possible. Mass advertising, especially through television, is able to eroticize commodities, and commodification whether or not specifically capitalist penetrates all areas of social life and all regions of the globe. Commodities become as never before status symbols, and the ability to purchase commodities is extended by debt expansion to a degree unheard of in earlier stages; and yet, for the majority of the world's people meeting subsistence needs is problematic.

This extremely brief sketch of some of the features of the realm of consumption in successive stages of capitalist development is meant to indicate that what takes place in the realm of consumption and how it is organized differs between one stage and the next. But stage theory is basically synchronic, outlining as it does typical abstract types associated with stage-specific forms of capital accumulation. My aim at the level of stage theory is to explore the most characteristic types of consumption as they are articulated with other regions of the economy and with extra-economic realms such as politics and ideology.

9.3 THE REALM OF CONSUMPTION AT THE LEVEL OF HISTORICAL ANALYSIS

Since the two more abstract levels of analysis are primarily synchronic, the main focus at this level must be the study of change. It is at this level that we can study the ever changing social construction of various consumer subject positions, how they relate to other subject positions, and how they empower or disempower those who take up or are taken up by them. Of particular importance would be the clarification of the social construction of needs, wants, and desires as they channel human energy towards coveting and consuming an array of commodities.[17] I will not try to even begin to map out the historical evolution of the realm of consumption in the capitalist epoch. Instead I want to indicate some of the dimensions of consumption and some of the problems in studying it at this most concrete level of analysis.

The first and greatest problem is delimiting what we mean by "consumption". In the context of pure capitalism, its meaning is quite clear and distinct, but at the level of historical analysis it becomes so potentially fecund as to defy efforts to fence in. Especially in this day and age when nearly everything either is a commodity or is treated like a commodity, it is not clear where consumption ends and other realms of life begin. We consume services. Do we consume the sexual services of our lovers? Do we consume the nurturing services of our parents? We consume leisure, but do we consume an afternoon nap? Do we only consume those things that we ultimately buy with money, or do we also consume things that come to us through some other kind of exchange, transaction, or expenditure of our life's energy. For example, when a friend returns a favour is the service performed by the friend consumed by me? Or when we consume a material commodity are we also consuming the environment? Do we "consume" the air we breath. The term "consumption" has been extended so far in the present period of history, that it is no longer clear what it means. Perhaps it says something about the stage of capitalism that we are in when almost our every act can be considered as an act of consumption.

Assuming that consumption is mainly of commodities or commodity-like things and that we more or less know what sorts of things are commodities, then we need to examine:

1. The range of things consumed as commodities, and how commodified they are.
2. How and by whom they are produced (capitalistically or not and under what conditions).
3. How they are consumed (especially who does what kind of labour in connection with consumption).
4. Who does the consuming and for what reasons.
5. How does the organization of consumption relate to struggles against exploitation and domination, or more broadly what kind of power relations prevail in the realm of consumption?

Between different periods of capitalist history, between different regions of the globe, and between different classes and strata there may be very different ranges of things consumed as commodities. A self-sufficient farmer may not consume any commodities. The starving poor may not consume enough to live. Also vastly different types of things may be consumed from bread to battleships, from woollen caps to hydro-electric dams, from shoes to transportation systems, from hovels to space stations.

The more technologically complex and expensive the item, the more government is likely to play a role. Also the more collective the consumption is, the more government is likely to play a role. Also the more developed capitalism becomes, the more the consumption of one thing is interdependent with the consumption of other things (Hirsch, 1976, 2). So, for example, the consumption of a car depends on roads. Finally the enjoyment of certain commodities by certain people may indirectly or directly effect the present or prospective enjoyment of other people or future generations (Hirsch, 1976, 4). For example, if too many people consume cars, the resulting gridlock may defeat the purpose of consuming cars in the first place. Or if this generation indulges in consumption that damages the environment, this will effect the consumption of future generations.

The issue of the extent to which various products are commodified is also an important issue to examine. In its purest form a commodity is capitalistically produced by a large number of producers for an impersonal market consisting of an indefinite number of consumers. Products that are bought and sold, but that do not fully meet these stringent criteria may still be commodity-like to varying extents. For example, military products that are produced on a cost-plus for the government have only some commodity-like features.

Commodities or commodity-like products may be produced capitalistically, or by self-employed labour, by domestic labour, by feudal labour, by slave labour, or by self-organized collective labour, to mention some of the more characteristic types. The production relations under which commodities are produced may or may not have an impact on their consumption. Domestic labour, which makes up the lions share of the labour of consumption, will directly impact on the organization of consumption. Also cheap labour, by cheapening the commodity, will expand its market. Finally, the more commodities are mass produced capitalistically, the more mass consumption becomes a crucial moment integrated into the profit-making circuit of capital.[18]

How are commodities consumed? Are they consumed individually, in families, or by larger groups or collectivities, or by the public as a whole? Different kinds of commodities are consumed more individually or more collectively. A transportation system (which typically would only be "commodity-like") is consumed collectively and a toothbrush individually. Also it is crucial to explore the kinds of labour associated with consumption including, for example, shopping, transporting, transforming, or organizing. A cake mix may be bought, transported home, cooked, and then served at a birthday party that one has organized. Each of these steps involves a different kind of labour.

In recent years more and more attention has focused on the question of who consumes for what reasons? The "who" need not be an individual, but could be a group or a legal person such as a state or a corporation. Consumption may be for many reasons. A few reasons are self-preservation, self-advancement, self-celebration, self-display, self-escape, self-destruction, or perhaps for entirely capricious reasons ("self" here may refer to a heterogeneous collective subject). And we have learned through modern advertising that commodities can be made attractive by appealing to the whole range of human emotions including most importantly erotic feelings.

One function of consumption is the gain in status that comes from the consumption of certain commodities. In some of his theorizing about consumption, Baudrillard (1981, 68) seems to have expanded this status function into a universal sign system that creates social order by distributing status. Such an approach tends to universalize and culturalize consumption with the result that production becomes reduced to a kind appendage of consumption, and it tends to blind us to the many and changing dimensions of consumption and production in all their historical specificity.

Finally, and most importantly for Marxian political economy, our analysis of consumption might explore how the organization of consumption plays a role in the development of class struggle and social movements. Some theorists have pointed out that we are all consumers and that therefore as consumers we constitute an undifferentiated mass (Ewen, 1976). Where consumer identity is stronger than worker identity, it follows that identification with the mass will be greater than identification with class. It would seem then that the realm of consumption would be a conservatising force pushing identities based on exploitation into the background. While there is no doubt something to this, a closer look might discover that there may be radicalizing forces at work in the realm of consumption. For example, in the stage of mercantilism, one of the most typical forms of class struggle was the bread riot. Bread riots were often led by women since they were the ones who usually bought the bread. One of the principal causes of bread riots was the development of a national, London-centered market which destroyed traditional local controls over bread prices. In the depression of the 1930s, the drastic lowering of workers' standards of living no doubt had a radicalizing affect. In the present period consumers have mobilized around a variety of causes usually having to do with the harm caused by specific commodities or with the abuse of power by giant corporations or by despotic governments. Sometimes consumer boycotts have been effective in bringing pressure to bear against specific corporations, industries, or even regimes. In the coming age of severe ecological crisis, consumers will increasingly organize against ecologically damaging commodities. Also with inequality, which appears to be growing on a global scale, workers in many advanced capitalist countries are facing serious erosion in their standards of living. In most large cities of the world, workers can no longer afford to buy single-family dwellings, and the day is perhaps not far off when they will no longer be able to afford cars.

9.4 CULTURALISM

There is an important "cultural" dimension to the realm of consumption and it is no doubt true that Marxists have tended to neglect this dimension.[19] At the same time, it should be stressed that "culture" is a notoriously vague concept that can be filled with diverse content. Culture can be about ways of life, attitudes, values, beliefs, emotive expression, cognitive expression, signs, symbols, discourses, ideologies, socialization, and material practices to name a few of the more prominent emphases in the use of the concept. "Culture" groups together such diverse and disparate phenomena under a common rubric that one might best consider cultural variables without using the concept "culture" at all. When one uses the term "culture" for analytic purposes, because of its vagueness, there is a need to specify some determinant meaning for the concept. The difficulty in clarifying the meaning of "culture" means that theoretical approaches that make it a central category have a considerable difficulty to overcome if they are not to be highly impressionistic.

Some recent approaches represent extreme forms of culturalism that come close to treating culture as the base and everything else as the superstructure. For example, McCracken (1990, 73) asserts: ". . . culture constitutes the world by supplying it with meaning". And for Sahlins (1976, 184), ". . . production is the substantialization of a cultural logic". In contrast, the levels of analysis approach that I have outlined here stresses the importance of the economic in understanding capitalism and it enables us to have a precise theory of capital's economic logic at one level, while considering how this logic articulates with "cultural" and other variables at more concrete levels of analysis. For the effective analysis of the realm of consumption, I believe it is crucial to be clear about capital's economic logic and about the specificity of capitalist economic variables. Among other things this means not following Baudrillard in converting commodities into signs, which has the effect of collapsing economics into culture.

I also have reservations about those approaches that make a theory of need central to understanding consumption. This is because "need" is a highly complex idea that is both socially constructed and unstable. For example, why does the typical Canadian worker "need" a car? The obvious and common sense answer is "to get around". But in 1900 the average Canadian worker did not need a car. Cars were relatively unusual luxury items owned only by the rich. They were not even very practical given the paucity and poor quality of roads. The story of how this new invention was gradually converted from a luxury item in 1900 to an everyday need of the average worker by the 1950s is long and complex. The technological developments that improved cars, cheapened them, and made them suitable for mass production played a role. The availability of cheap petroleum, and of many other important inputs in the construction and operation of cars played a role. Geo-political factors such as suburbanization and highway construction played a role. The political clout of auto corporations and the neglect of alternative transportation systems played a role. The expansion of debt financing played an important role. The development of mass media and advertising played a role. And then there are specific characteristics of the commodity as it was produced in the 1950s and 1960s such as built-in obsolescence, frequent model changes, and strong emphasis on status differentiation. In short, it is not useful to think of "need" as some sort of psychological variable that arises spontaneously within individual subjects or is constructed by something so vague and general as "culture." Instead the object of the need, the subject of the need, and the need itself are the results of highly complex processes of socio-economic construction that extend indefinitely into the past. This means that there can be no very useful general theory of need, since needs are always complex constructions located within specific socio-historical contexts. Because Baudrillard apprehends this and lacks an idea of levels of analysis, the concrete is made imperialistic over abstract theory. The law of value is defeated because of the contextual construction of needs by an encoded sign system.

In the above list of factors involved in the social construction of the need for a car, economic factors appear to play a more important role than cultural factors. Indeed, the mass production and mass consumption of automobiles has probably had a much greater impact on culture than the other way around. Cultural considerations, or what I would prefer to call "ideological" considerations, mainly help to explain things like the intensity of one's lust for a car, or why one prefers a muscle car to a sub-compact, or frequency with which one changes cars. The basic fact that the average industrial worker in Canada today "needs" a car requires explanation primarily by economic factors and only secondarily by cultural factors.

Not only do culturalist approaches fail to grasp the specific weightiness of economic variables in explaining something like the social construction of the need for a car, but some of them by their particular emphases may compound their one-dimensionality. For example, to generalize from the clothing fashion system a theory that converts the commodity into being one sign amongst others, as Baudrillard does, is bound to produce serious one-dimensionality. First of all, to the majority of people in the world who live on the edge of bare physical survival, the fashion system cannot be very relevant. Second, where the fashion system is relevant, the status function of commodities, is only one function and often not the most important function. For example, are we to understand the social construction of the need for workers to buy cars as nothing more than the arbitrary creation of a fashion code?

The purpose of the above considerations is to bring out the difficulty in theorizing the reasons, motives, and needs for consumption, and to indicate the pitfalls in culturalist approaches and approaches that place a general theory of needs center stage. Indeed, referring back to an earlier quote from Sahlins, it is not obvious to me that reasons why in the main men (approximately one male in a hundred in Canada is a cross-dresser) do not consume skirts (except in Scotland and certain other countries) is an issue that should only be approached as a question of culture, or that this rather trivial example should be generalized to support a universal culturalist approach to the understanding of consumption. Even something so seemingly arbitrary as fashion may be shaped and constrained fundamentally by economic variables and perhaps even more by political variables if we include amongst them power relations. Indeed, fashion is totally laden with the power that adheres to various subject positions.

9.5 CONCLUSIONS

The glitter of the commodified surface of everyday life makes it easy for us to become both bemused and enchanted by the swirl of flashy surfaces around us. Some trends in postmodern thought seem to be quite caught up in this enchanted world of labels and aestheticized commodities. They turn the fleeting surfaces of commodities into an epistemological credo, that makes a virtue out of being concerned only with the surfaces and fragments of the world of glitter, packaging, facade, and immediate flux.

I have presented a levels of analysis version of Marxian political economy that stands in sharp contrast to Baudrillard's fixations on the commodity as an element in a general sign/fashion system. I have asserted that capital has an inner logic, and that knowing this logic gives us important insights into capital's identity and mode of operation. Also I have argued that capital has passed through qualitatively distinct stages where what consumption is and how it is organized alters significantly. Finally I have explored some of the dimensions of variation in the realm of consumption and some of the pitfalls in studying it at the level of historical analysis.

In the context of pure capitalism, the realm of production is primary because that is where profits come from and because profit-making is the overriding end of capitalism. In so far as profit-making is still central to capitalism, we cannot ignore the importance of the realm of production, even though in the current stage of capitalist development profit-making is organized in such a way that arguably consumption may be considered in many ways nearly co-equal with production in importance, particularly when the ideological and political dimensions of consumption are considered. The problem with Marxist theory that Baudrillard and others have reacted so strongly against is the tendency to apply Marx's laws of motion of capital directly to concrete history. But the levels of analysis approach avoids this by maintaining the laws of motion at one level, while carefully bringing in complex mediations at other levels. Thus, even at the level of stage theory the realm of production cannot be given primacy in the same sense as it is in the context of pure capitalism. Indeed, in the present stage of capitalist development, production and consumption are so enmeshed with each other that it is probably not possible to determine which is in some sense more primary or more determinant.

Understanding that the realm of consumption is secondary and passive within capital's inner logic and understanding why it is can give us insights into many issues. For example, the centrality of production at least partially explains why domestic labour has been devalued in capitalist societies. Domestic labour is not valued because it is in the realm of consumption and does not therefore directly produce surplus value. It is a fact of life of capitalism that the discipline of economics is the science of the "main business", while the devalued science of home economics is the science of domestic labour (Smith, 1989). This devaluing of various forms of the labour of consumption cannot be altered by a simple conceptual reversal that makes consumption primary. It is capitalism itself that must be altered, and the more thoroughly we can trace all the threads of power that enmesh us, the more we can effectively unravel them. Real consumer sovereignty has yet to be seen in the world, and it is unlikely to appear as long as capitalism remains.

I have claimed in this paper that while Marxian political economists may have neglected the realm of consumption because they have applied Marx's *Capital* too directly to history, that a levels of analysis approach can overcome this problem by avoiding reductionism and introducing the necessary complexity into the theory. I believe that the approach I have outlined can build upon the tremendous explanatory power of Marx's *Capital*, while helping us to explore consumption in all its dimensions, as we move from abstract economic law towards the immense complexity of the concrete. In order to understand consumption in a capitalist society, it is essential to understand capital itself – what it is and how it operates. The need is to deepen our understanding of capitalist economics by a fuller exploration of its "cultural" or ideological dimensions and not to collapse the economic into the cultural thus losing the materialist grounding of Marxian political economy and the powerful and penetrating analysis of capitalism bequeathed to us by Marx and Marxism.

Notes and References

1. The work of Baudrillard (1975; 1981) is currently the most fashionable and influential. For interesting commentary on Baudrillard see Kroker and Kroker (1991). For an effort to build on the realm of consumption from the perspective of cultural anthropology see M. Sahlins (1976). For an effort to give the realm of consumption more theoretical weight from an Hegelian perspective see R. Winfield (1988). For an interesting and influential analysis of consumption from the point of view of limits to growth see F. Hirsch (1976); and also see J. Kassiola (1990). Two efforts to show the importance of the realm of consumption in early capitalism are J. Brewer, N. McKendrick, and J.H. Plumb (1982), and M. Berg (1985). For overviews of recent work on consumption see M. Featherstone (1991) and G. McCracken (1990).
2. I have called this tendency "the logical-historical method". For an extended critique of the logical-historical method see Albritton (1986).
3. It may seem as expressed here that production and consumption represent a binary opposition in need of deconstructing in Derrida's sense. However, in this case the binary opposition is not a product of our logic but of capital's logic, and hence can only be deconstructed by deconstructing capitalism itself (i.e. by socialism).
4. "An object is not an object of consumption unless it is released from its psychic determinations as *symbol*; from its functional determinations as *instrument*; from its commercial determinations as *product*; and is thus *liberated as a sign* to be recaptured by the formal logic of fashion, i.e., by the logic of differentiation" (Baudrillard, 1981, p. 67, italics in the original). Furthermore, "The sign no longer designates anything at all. It approaches its true structural limit which is to refer back only to other signs" (Baudrillard, 1975, p. 128).
5. "What is consumed is not a thing . . . but purely and simply an element in a code" (Baudrillard, 1975, p. 9).
6. A systematic critique of Baudrillard would be very difficult because he is not a systematic thinker.

7. See Uno (1980). See also T. Sekine (1984; 1986) and Albritton (1986; 1991).
8. I have considerably expanded on the work of Uno and Sekine, particularly in the specification of stage theory. One significant innovation is that Uno theorized three stages: mercantilism, liberalism, and imperialism; wheras I add a fourth stage: consumerism.
9. See Albritton (1991) for an expanded treatment of levels of analysis, particularly mid-range theory, or what I call, following Uno and Sekine, "stage theory". For a brief outline of the epistemology associated with levels of analysis see Albritton (1992).
10. ". . . [N]eeds erect themselves more and more into an abstract system, regulated by a principal of equivalence and general combinative" (Baudrillard, 1981, p. 135). ". . . [T]here are only needs because the system needs them" (Baudrillard, 1981, p. 82).
11. The distinction between productive and unproductive consumption is only clear and precise at the level of pure capitalism where only those commodities consumed directly in the process of capitalistic production are considered to belong to the category "productive consumption". At more concrete levels of analysis there may be grey areas that could occasion considerable debate. While being aware of these grey areas, in this paper when I use the term "consumption", I mean unproductive consumption unless otherwise specified.
12. Baudrillard (1981, 63, 71) seems to think that Marx theorized consumption as a relation between a subject, an object, and a need, and that this conceptualization needs deconstructing.
13. Marx writes: "With the development of capitalist production, the scale of production is determined less and less by the direct demand for the product and more and more by the amount of capital available in the hands of the individual capitalist, by the urge for self-expansion inherent in his capital and by the need of continuity and expansion of the process of production" (K. Marx, *Capital* Vol. II (Moscow: Progress, 1971), p. 147).
14. For an interesting application of the sequence textiles, steel, and autos see J.R. Kurth's (1979) use of product cycle theory to analyze the political outcomes associated with various articulations of this sequence.
15. The capitalist production relations most typical of the stage of mercantilism developed in putting-out manufacturing in England between approximately 1700 and 1750. Most historians estimate that putting-out workers spent on average 70% of their income on food and drink.
16. Domestic labour has generally been the largest category within the larger category labour of consumption because most consumption has taken place in the domestic setting. However, there are also various categories of collective consumption that occur outside the home such as the consumption of various forms of recreation or transportation systems.
17. It is at this level of analysis that some of Baudrillard's ideas become most useful.
18. It is particularly "Regulation Theory" that has based much of their perspective on the importance of mass consumption in the "fordist" epoch. While I agree that the realm of consumption attains an added

weightiness after World War II, I disagree with the way in which Regulation Theory theorizes the realm of consumption. For elaboration, see my article in this collection.

19. I am uneasy about using the concept "culture" because of its vague and all-encompassing character.

10 A Very Nearly Capitalist Society

Colin Duncan

The case I wish to advance here involves the use of a peculiar and I believe novel kind of theoretical argument on an unsuspecting body of historical research which was itself developed mostly by the use of purely empiricist methods. In consequence I think that, although it may at first seem inappropriately autobiographical, it will be helpful if I begin by recounting how I came to develop the idea I intend to describe.

The process was not a rapid one; it being now a little over eleven years since I began trying to understand Kozo Uno's work. A year after it came out in English, I read Sekine's translation of Uno's *Principles of Political Economy* on my own,[1] but almost immediately after I arrived in Toronto to begin postgraduate studies at York University, I was lucky enough to be able to embark on the study of political economy under the formal tutelage of Professor Sekine, who was at the time nearing the completion of his massive *Dialectic of Capital*.[2] Since the latter work can be regarded in some respects as a detailed commentary upon Uno's key concept of the theory of a purely capitalist society, and constituted the set text for Professor Sekine's year long course, I received a fairly thorough exposure to questions of theory and method at that time.

My previous exposure to these issues had been rather conventional: a reading course on Volume One of Marx's *Capital* four years previously under the direction of a materialist philosopher of science who was greatly interested in the bearing on the methodology of social explanation of the general philosophical distinction between appearance and reality. I cannot recall precisely what I thought about the relation of social theory to historical reality back then, but I am fairly sure that prior to studying with Professor Sekine I must have been rather naively thinking that *Das Kapital* was somehow a book about how England's economic system really worked, the appearances paraded in bourgeois ideology notwithstanding.

Of course while studying Professor Sekine's *Dialectic* I came to see that concrete history can hardly (and only in a few places at that) have been very closely homologous to the theoretical model described in political economy. Early in the year the members of the class were advised by Professor Sekine to read Karl Polanyi's *The Great Transformation*[3] (1944) with care. I immediately saw from that text that there was also an explicitly political component to the process of emergence of capitalism in England, and that detailed attention has to be paid to the early 19th century in particular, for it was only then that the notion attained wide currency of confining all the modes of livelihood of a whole society within a single integrated system of price-setting markets. Polanyi however did not make as clear as Professor Sekine the further, implicit, and culturally more ominous point that it was also thereby being suggested that all manners of livelihood were henceforth to be subjected to the requirements of a single, secular abstract idea: the self-expansion of value.

Now while Uno at times spoke as though he thought the theory of a purely capitalist society is a theory about the deep reality of actual capitalist societies, and while he also seems to have been sure that the "fit" between theory and reality (to the extent it was actual) was very much a result of much deliberate state policy,[4] it is hardly the case that either of these rather vague and yet familiar points can help much to elucidate the detailed peculiarities of Uno's conception of the relation between history and theory. I must confess that at first then I had very little idea quite how to elucidate (let alone digest) the implications for historical understanding of Uno's distinctive position.

Happily for me, however, in my first year at York I was also able to study with Sekine's colleague, Professor Albritton, who was that year for the first time offering a course on imperialism, the highest stage of capitalism. We spent much time in that seminar trying to elucidate Uno's concept of stages in capitalist development, and indeed the two books Professor Albritton has written since then testify amply to the extent of his interest in that topic.[5] If I may presume summarily to formulate Professor Albritton's problem as it was seen at that time, it was how can we get from the theory of a purely capitalist society to stage theory. It is important to stress that this was seen as a theoretical problem, that it was thus taken for granted that the direction of inquiry was, as it were, downwards. Apparently Uno did not get around to laying out very systematically what he thought about stage theory (apart from his organized comments about the history of state policy) and so there has been a certain amount of groping by his students in this area. There seemed to be general agreement however that once stage theory had been worked out it would be found to be useful to historians, certainly much more directly useful than the theory of a purely capitalist society.

For myself there the matter stood for a while because the very summer after my first academic year at York I took another course which also influenced my thinking, but which was utterly different in approach. It was a brand new methodological course on social history taught by an eminent social historian of England, Professor Nicholas Rogers. In that seminar I was exposed to the ways of thinking of historians deeply interested in the "blooming, buzzing confusion" of social life in all its varied and concrete detail. This experience made a lasting impression on me. Indeed ultimately it led me to attempt an unqualified jump from the interdisciplinary rapids of social and political thought onto the disciplinary island of history. That I have now spent more than half a decade on that island makes me bold enough to try to argue for an approach to the problem of stage theory from, as it were, the diametrically opposite corner to Professor Albritton's.[6]

10.1 DOING HISTORY "UPWARDS"

I realize I could instead have chosen to speak of my approach as doing stage theory "from below", but that debased phrase is only too apt to suggest merely some crude politicization of the issues involved – a way of proceeding very antithetical to Uno/Sekine Riron. As I will explain political enthusiasms did not lead me by the nose in this matter. On the contrary my point of departure has been a tantalizing metahistorical allusion to the contingent complexity of concrete human life made by Professor Sekine himself:

> The logic of the commodity-economy can only synthesize a purely capitalist society in which use-values are totally subsumed under the form of value, not any concrete capitalist society that exists in history. The latter more or less deviates from the former depending on the use-value character of real economic life which is by and large an historical product.[7]

Since this passage occurs in the middle of an explanation of what was special about Uno's theory of imperialism, I first understood it to be alluding to the point that one could hardly expect the society that produced commercial steamships, for example, to operate with the same elasticity as one in which there was very little in the way of such capital-intensive production (outside of the state sphere of ordnance at least).

The point made in the quotation is a more general and indeed extraordinarily powerful one however. It immediately appealed to my materialist sensibilities, also to my common sense, which had long, albeit inchoately, been rebelling against the great lie of commodity society, namely that all economic goods might as well be thought to have identical physical characteristics, and that it therefore cannot make any important difference what specific goods are wanted during any period. Indeed of all Professor Sekine's many profound but unfortunately analectical formulations, this one I just quoted has stuck most in my mind. For me the clause "depending on the use-value character of real economic life which is by and large an historical product" has assumed something of the character of a mantra. What I wish to suggest here is that it may be the key to writing the social history of capitalism in a novel manner that could be comprehensive and yet non-arbitrary.[8]

10.2 CONFRONTING THE EXTENT OF COMMODIFICATION SERIOUSLY AS A PARTLY INDEPENDENT VARIABLE

What I am envisaging would be a rather systematic foundation[9] for a history of modern life based firmly in attempts to answer the following three related and fundamental complex questions:

A. What goods were actually wanted, and how were they produced, and in what quantities, during every stage in the "real economic life" of a given society?
B. For each good, how amenable was it at every stage to the commodity form, both as regards the retailing and the production side; that is, to what extent could the market for it actually measure its (necessarily social) value?
C. What proportion of each good classified thus according to such an index of 'commodifiability' in turn actually took the commodity form at every stage?

I submit that only once a socio-economic history has been written which contains the answers to such questions will we have a history cast in the right form for the purpose of explaining the peculiarities of history under capitalism.

I also submit that it is only for the case of capitalist history that such a foundation for social historical study would be the correct one. In the case of other social forms an ancillary study of the commodified sphere would always be enlightening but it would be inappropriate to attach great importance to it, especially initially. In pre-capitalist societies "the use-value character of (their) real economic life" might or might not be very important as determinants of their fundamental character. It seems that in many societies genealogy, for example, or some form of religion, was the crucial unifying project. While it would be absurd to count them as epiphenomena, the activities directly embroiled in the pursuit of the material necessities of life in such cases must have seemed by implicit comparison to be merely instrumental. Indeed it is probably only in a society where the arrangement of the modes of livelihood had come to have a uniform, homogeneous character that chrematistics could possibly loom so large as to displace such complex human enterprises as the reproduction of kinship systems or organized religion. Certainly it is no easy matter to explain how means can have come to pose as ends. For, after the relevant empirical facts about goods have been gathered for the early modern and modern periods, the great interpretative trick will be making all our "basic" information about commodities speak, as it were, to the enormous and utterly chaotic body of cultural knowledge we already possess about ideologies and government policies, both during the development and maturation of capitalism.[10] I suspect that some of the details in the history of the commodification of the elements of livelihoods may help explain some of the changes in ideas and attitudes associated with modern history.

One way of putting the puzzle is to note that although capitalism is inherently a totalizing approach to social affairs which brooks no compromises and which was deeply alien in many ways to the institutions in which most people had hitherto lived their lives, yet in the case of England the capitalist way of organizing affairs came upon society, as it were, unawares, by capturing precisely the sphere of the ordinary daily life of the common people. Obviously English culture had to be somehow predisposed, by accident presumably, to allow such an "invasion", but my point here is that nothing very remarkable would have happened had not a large proportion of the use-values of the time, and the associated techniques for their production, been intrinsically amenable to a commodification process. It will be emphasized that the commodification of daily life is an inherently social process, much more complicated to explain in some ways than the associated changes in human motives often postulated to have been crucial in the story of the rise of capitalism.[11]

My aim then is to contribute to producing an historical depiction where the canvas is the story of the commodification of use-values onto which all the other themes of social and political history can be embroidered. I think that such would result in a more secure foundation for theoretical discussion of the stages of capitalist development. Currently stage theory has had perforce to rely upon histories written from all sorts of other perspectives. In this connexion it would surely not be too much to say that Professor Albritton has been engaged more or less single-handed in a massive campaign of historiographic pillaging. Considered in one light this has been very successful for he has thereby come to make a large number of very interesting points about the development of capitalism in England. But I fear that his results have been both more arbitrary and less reliable than they ideally could and (I am sure) eventually will be, since the underlying principles of selection in the works he has pillaged were at least foreign to our own, often patently incompatible, and of course frequently either unknown, or unknowable because so internally incoherent. What has given Professor Albritton's work the elegance and unity it manifestly has, has been his resolute focus on the sphere of the legal-ideological, the theory of the capitalist state, if you like. This has been a largely "top-down" approach and I am suggesting it needs to be complemented by studies oriented around the implications for commodity ideology of the real material heterogeneity of the use-values people actually sought. I do not mean to begin here the detailed and systematic task of listing goods and classifying them according to their compatibility with commodity-logic. Rather, having I hope sown the seed of the general idea, I wish now to consider a few of its grosser implications. Some of these relate to other themes in history which are well-known in their own right and which many others before have tried to link to the history of capitalism, but which assume a rather different and I think freshly interesting aspect when considered as auxiliary to the project I have sketched here. My commodity-based way of approaching capitalist history also obviously has implications for some old established problems in social theory. I will start by making a few remarks on what most would still call "the problem of how to classify different economic systems".

10.3 DIFFERENT TYPES OF ECONOMIC INSTITUTIONS AND THEIR VARIOUS INTERACTIONS

If we follow Karl Polanyi and distinguish the substantive sense of "economic" from the formal sense associated with modern economic theory, then we thereby isolate for separate consideration the notion of economic activity considered merely as "bearing reference to the process of satisfying material wants",[12] that is, with the institutional form left utterly unspecified. With the option having been given thus to discuss both the physical activity and the institutional forms in the abstract it becomes possible, indeed necessary, to set about trying to classify the latter. Now it would be the mark of a "totalizing" social theorist of the kind despised by the current self-styled avant-garde in social theory to imagine that these all shared some essence,[13] and indeed I would like to make a point of asking you not to consider the forms I will discuss as though they all occupied points on a single line or spectrum. The many divers social forms are distributed over a type of 'space' with many dimensions, certainly far more than three. Nonetheless certain broad structural congeries are discernible within this non-Cartesian universe of discourse. Three of these "clusterings" of institutional elements are what I will call: state dominated, peasant household, and capitalist forms of social being. I wish to make a few remarks about each in turn, but let me first suggest that an important difference among these social forms concerns how they vary in the extent of the range of human projects and purposes they respectively allow to their members.

State dominated social forms are typically utterly obsessed with some quintessentially unitary purpose: security, expansion for its own sake, the setting up of a dynasty, etc. A large fraction of substantive economic activity under such conditions is harnessed to that project. Certain goods are needed in certain quantities in order to ensure success. As for the rest of substantive economic activity the state often hardly cares how it is socially instituted and seeks only to minimize any political disturbance resulting therefrom.[14] Indeed in history many powerful states have had explicit arrangements with peasantries, the members of which often did not at all share, and sometimes even just barely tolerated, the particular (indeed often thoroughly private) obsession(s) of the ruler(s). You will I hope have noticed that I did not use the term "peasantry" in my short list of general forms. Instead I used the adjectival construction: "peasant household". This is because I believe that at the highest level of generality what is remarkable about what many carelessly term "peasant society" is that it is only presumptively called a "society" at all.

The category of peasant household in my usage refers to a group of social phenomena that may or may not constitute a social form as such. The kernel of the peasant outlook is a very clear focus on the survival of the household. It actually often resembles the state writ tiny! If in a given area there are other peasant households then from the point of view of any one the other ones are indeed apt to be seen as relevant parts of the social environment, but it is always dangerous to assume that peasantries naturally cohere enough to act together. The integrity of the household is always the first priority.[15] A peasant household typically occupies a particular part of the natural environment and proceeds to interact with it so as to procure a flow of certain basic goods. More often than not peasants are keenly aware that tomorrow follows today, that the seasons follow each other, and that times of dearth and plenty jostle each other successively as well. This serious apprehension of mortality coupled generally with deep awareness of the crucial details of their immediate natural environment has made peasantries collectively marked by a concern with achieving a sustainable flow of the set of goods they need, whenever they have been allowed to do so. I do not think the great Russian rural sociologist, Chayanov, picked up in this way on the peasant hallmark of sustainable production,[16] but he did usefully draw attention to another distinctive characteristic of the peasant household which is of great relevance in the present context. The peasant household has essentially finite wants. There is a set of goods needed to attain a comfortable life. What exactly the latter looks like of course varies historically, but the point Chayanov insisted on is that peasants toil until they achieve their life-style, whereupon they typically simply stop. Depending on the terrain they inhabit and the social environment, they may find it expedient to produce more of a good with which they are unusually well supplied and trade it away for quantities of some other good or goods. This indulgence in trade typically remains epiphenomenal however. The goal, the procurement of a particular set of goods in specified proportions remains the same. The basket of goods aimed at is generally a simple function of the terrain and the level of comfort sought. Of course the contents of the basket may change with time, usually rather slowly, but the goal of economic activity is inherently basket defined. All the particular goods must be there in the right proportions at the end of the day, or season, more properly. Indifference to the use-values of the goods in question remains superficial because substitution is exceptional and always self-consciously instrumental in any case.

The contrast with the situation of the household under ramifying capitalist conditions is crucial. Of course every household under capitalism has at any point in time a finite budget and aims at a certain basket of necessities-cum-desirables. To that extent it thus superficially seems to be in a condition similar to that of the peasant household. But whereas the inherently variegated purposes of the peasant household remain essentially finite those of the capitalist household are always also vulnerable to the imperious dictate of another purpose, a deeply abstract and mysterious one that everyone who lives under capitalist conditions nevertheless knows strides abroad by day and night. I speak of course of the self-expansion of value. To the extent that a member or members of a household under capitalism become the servant(s) of chrematistics, the orientation of substantive economic activity shifts fundamentally away from a preoccupation with usefulness and comfort as such. It also systematically tends to push the concern with sustainability to the background, but that is another story.[17]

By means of the above discussion of substantive economic activity I hope to have underscored the deep peculiarity in terms of daily life of capitalist indifference to use-value. When committed economic liberals profess that they are baffled by the "narrow" economic outlooks of the state on the one hand and the peasants on the other, they are registering the existence now of a profound dichotomy in human affairs. Either one views material goods as inherently tied to their uses (ordnance for defending the monarch's family, roof tiles for keeping out the rain, root crops to store over the winter months, etc.) or, to put it crudely, one doesn't. Even today people more or less uneasily acknowledge that there is something odd about the merchant point of view, whereby one buys only in order to sell, and makes a point, as it were, of being indifferent to the useful qualities of the good. To give another example, it remains difficult for many people to see the so-called "opportunity cost" style of argument as decisive when the question is whether to sell a beloved home that merely happens to have become surrounded by 'yuppies' who have driven the so-called "market value" up merely by invading in large numbers. I will conclude this section by proposing that the proliferation of the merchant outlook that characterizes life under capitalism probably had to be predicated on the attainment of relative abundance. Until most householders became convinced through experience not only that they had very little chance directly to affect the likelihood of their managing to get all the things they needed, and that there was in any case only a very low probability of a shortfall of supply within their lifetime of any of the key elements in their baskets, could they even consider embracing the abstraction, Value, as a god. Of course, as is notorious in our century, once any significant number of persons with idle funds become the servants of Value they develop a strong interest in accelerating any commodification processes already extant. The conditions for the emergence of this class of social "accelerators" are also thus relevant to our overall project.

10.4 REAL CONSTRAINTS ON ECONOMIC ACTIVITIES AND THEIR UNINTENDED CONSEQUENCES INCLUDING COUNTER-TENDENCIES TO COMMODIFICATION

In the discussion so far the whole strategic point has been to consider substantive economic activity and its contingent institutional garb in the abstract. Let me now turn to consider the set of general constraints on the patterns that economic activity can take. These constraints can be considered under a number of heads but for now I will only resort to illustrations which help to clarify the English process of undergoing commodification. The survey that follows will thus not be comprehensive, let alone exhaustive. I may alert you to look out in advance for the interesting circumstance of what seems a constraint from one point of view actually looking like an opportunity from another, and vice versa. Human history is positively littered with cases of people accepting some particular development on one set of grounds, even though its tendency was actually to alter circumstances in general, and thus the whole basis of their original decision. Most liberal ex post facto justifications of the status quo, that is, most works of economic history, actually exhibit the logical fallacy of irrelevance, pointing to ends not considered at the time the means were selected. When challenged on this score the offenders of course usually compound their illogicality by gesturing to what they call "Progress" in order to end discussion. Such sloppy intellectual habits impede the clear understanding of what capitalism, as distinguished, for instance, from industrialism implies for social life.

First let us consider the implications of geographical factors for the study of how the commodity form can overrun pre-existing patterns of livelihood. Beside being an island in the path of moist prevailing winds of generally moderate temperature, one of the features which distinguish the English landscape from many, though not all, of the landscapes of other European countries is the large expanse of gentle terrain which allows for a relatively even distribution of the generous precipitation. In addition to being an agronomically unifying factor which encouraged a relatively even distribution of population this pattern of topography and rain gave England a large network of slow and hence navigable rivers which encouraged trade outside one's immediate locale. England developed a large network of very small market towns with significantly only one large city, London.[18] The latter in addition to being England's only urban excrescence or "wen" as Cobbett called it, was by far the largest European city in the modern era. Within London then the division of labour could proceed very far as highly specialized markets could be served, both markets in unusual goods and markets in parts of goods. This circumstance favoured technical change in the English economy, both as regards the production process and the retail system. For in the rest of the country the network of small shops was and remains remarkable. Eighteenth century England saw the attainment of the highest ratio of shops per capita seen on Earth as yet.[19] Napoleon's derisive phrase, "nation of shopkeepers", fit the case surprisingly precisely. Significantly a very large proportion of these were rural shops. Admittedly the standard range of commodities on offer in them was very small, especially at first (it centred on light, cheap imports: chiefly tea and sugar), but it is undeniable that the socio-institutional effect of this early ramifying commerce must have been considerable.

It is worth noting here that England's subsequent patterns of population distribution associated with the rapid growth of a myriad new cities in the early and mid-nineteenth century probably were less important in facilitating the spread of commodity relations. It must be remembered that the industrialization that occurred in England's new cities-to-be was entirely predicated on an equally remarkable though as yet much less remarked upon de-industrialization of the countryside. The late Victorian image of the purely agricultural country as opposed to the purely industrial urban areas depended on a very recent development.[20] Adam Smith would have been very surprised at, and I don't doubt troubled by, the irrationally hyper-urbanized version of capitalism that England came to represent over the course of the 19th century.[21] The people of the erstwhile Soviet Union are not the only victims of the collective amnesia which has enabled us to "forget" that the connexion between capitalism and extreme urbanization is purely adventitious.[22] The finishing touches on England's urbanization process have of course been due to the chronic depression in arable farming which has been with us since the 1870s (apart from a brief interlude of good times surrounding the First World War).[23] Of course the hyper-urbanization in late 19th century England and indeed in Western Europe as a whole as well, called forth (and soon came to depend upon) a far-reaching commodification process with respect to foodstuffs from the prairies of the world. That is a story we cannot go into now, but I find I cannot refrain from remarking here that to the extent that a putatively capitalist society (such as England ca.1900) relies heavily for basic goods on less capitalist regions overseas, it cannot be simply described as a capitalist society, as such. It remains a capitalist social formation perhaps but surely the idea of a society whose members base their diet on agricultural goods but which has expelled its own agriculture is a sociological fraud at one level at least. This is a point that probably would never occur to any of the self-styled "post-modern" theorists but it is one that has worried military authorities since the heyday of Ancient Greece. The context of the above discussion on food supplies brings up another aspect of hyper-urbanization which is worth mentioning because it allows us to raise the general question of counter-trends to the commodification of daily life.

Hyper-urbanization leads to a generalized increase in the costs per capita of sustaining life. As Leopold Kohr made extremely clear half a century ago there is a direct relationship between social costs and social scale.[24] The costs he drew attention to are not randomly generated. They cluster heavily around policing, sewage, drinking water supplies, fuel supplies and transportation. It is well-known that many of these areas of concern to the citizens of hyper-urban areas have come to be dealt with on a municipal basis. The supplying of the associated seeming "needs" (which would either not exist or not be so imperious if it were not for the hyper-urban condition itself) has frequently taken a form either not amenable, or less amenable to commodity logic, and has not infrequently taken an explicitly anti-commodity form. At one point it was even thought that such "municipal socialism" would provide the building blocks for fully socialist national economies. It could be argued then that a social form aspiring to become and remain a purely capitalist society would have somehow to keep its urbanization process within limits. Beyond a certain point urbanization and capitalism are mutually contradictory.

10.5 IMPLICATIONS FOR SOME PROBLEMS IN THE HISTORY OF IDEOLOGY

I now wish to advance yet another rather unfashionable line of argument. I believe it is almost impossible to overestimate (although it would be even harder to quantify!) the role of the Reformation in capitalist development. But my point is not the usual one about individualism and the bourgeois self, although there clearly is a lot in that old argument.[25] Rather I wish to suggest, following a typically fertile lead from Professor Sekine, that Nietzsche was wrong about the date of God's death. God effectively ceased to be when He ceased to be one. As I now see it, the Reformation made two Gods and two Christs where before there had been only one of each. This point occurred to me when reflecting upon a style of invective indulged in during the time of the English Civil War. It was common for both sides to accuse the other of representing anti-Christ. It struck me that it would be quite impossible for us today to come up with a correspondingly totalistic condemnation. When both sides believe there is only one correct religion, and no third or "external" position is allowed, then such an absolutely damning charge can nowise be evaded. But the long-term harvest of this period of turmoil, however paradoxically, was increased toleration, even of atheism. Our world came in fact to regard religious belief as just part of one's moral-intellectual make-up (optional at that), and thus we can no longer hurl correspondingly overwhelming abuse. This simple example from the parlance of the day underscores the deep socio-psychological ramifications of the Reformation which I think help to explain how Value could even come to be considered as a rival to the supposedly unitary Christian god. Much work needs to be done in this area, but I think we can say that the rather simple-minded focus so far on "individualism" as the key or only relevant result of the Reformation has been exaggerated. It has been made abundantly clear by much new work on the 19th century (and indeed the later 18th too) that the antinomy of Catholic social church versus Protestant individualistic anomie is too coarse an anaytical tool. Davidoff and Hall have indicated for example that the English upper and middle classes made the nuclear family into their minimal religious unit. This was particularly the case for the "middling sorts" who embraced evangelical religions so fervently during that most crucial post-Speenhamland period that forms the centre-piece of Karl Polanyi's *Great Transformation*. Again it is the household perspective that could give us many new insights, I would like to suggest.[26]

Religious upheaval was not the only sort of change that had a relevant unsettling effect on the ideology of the English during the development of capitalism. The Civil War was also a fight about the constitution. The fact of debate about the right of monarches to rule absolutely, and the nature of the eventual 'solution' called the Glorious Revolution, gave rise to a continuing debate about the extent to which private families could be expected to simply accept the whims of the person who merely happened to be the monarch of the day. Eighteenth century politics is perhaps best described as a dynamic equilibrium with tension apt to flare up in the form of intense political factionalism at the slightest provocation.[27] It is hard to view the contentious political arena and not feel it must also have had a profoundly destabilizing ideological effect. Surely again the seductive abstraction Value was the ultimate beneficiary of this internecine ideological wrangling amongst the English. The mere fact of all the religious and constitutional conflict of the 17th and 18th centuries must have been propitious for new developments in general. In fact the particular outcome of constitutional conflict, whereby the strong common law tradition from the past defeated the absolutest pretensions of monarchy and allowed the emergence of an extraordinarily powerful and forward-looking aristocracy, directly helped the spread of capitalism by its effect upon commercial agriculture.

The great landowners of England, a group which had swollen due to the Henrician bonanza more politely known as "the dissolution of the monasteries", came to feel they had a strong interest in establishing their families upon the most secure basis possible. This took the form of heavy investment in land. As a class overwhelmingly dependent upon rent they had a manifest interest in seeing that the productivity of their land was increased, minimally that it did not decrease. The English landowning class which was also the ruling class saw to it that the law positively facilitated the development of agriculture. Since the rents they charged were substantial, and thoroughly monetized, the stimulus to commercial farming was very significant. A better, more secure basis for the emergence of capitalist social forms could hardly be imagined. Adam Smith made explicit comments to that effect in his famous Inquiry.[28] It needs to be emphasized that the importance of this is apt to be seriously downplayed by persons from the 20th century. Until well into the 2nd half of the 19th century it is a simple fact that, apart from rent, by far the largest category in the budgets of the vast bulk of the population was a single item, of agricultural provenance: bread. Together rent and bread accounted for about a half of most working class budgets.[29] It is worth noting then that while on the one hand the list of commodities sold in mass quantities on a sustained (that is, regularly repeated) basis in the English domestic market remained very short in absolute terms for a long time, on the other hand the proportion of the non-rent part of household budgets spent on fully commodified goods was virtually 100%. The commodified sphere was all that most people knew! It has probably never again been so relatively pre-eminent. This raises a point which can serve well as the final one in this outline of a new research project.

10.6 CONCLUDING REMARKS

Overall, in the plan for a new history of capitalism which I have been roughing out here, we unquestionably see the commodified area expanding relentlessly, even inexorably, but the fact that the "story" seems to complete itself must not blind us to the extreme peculiarity of this process. For instance, it becomes an interesting question whether the sequence whereby all the various different goods were commodified was in itself significant. Certainly we cannot look to the economic theory of capital for guidance on this score precisely because, from its use-value indifferent point of view, there can be no prioritizing of commodities. None is more important than any other. Capital itself has never allowed any set of distinctions to get a grip in our ideology. In that respect it is utterly the opposite of the peasant household outlook. Nonetheless it is worth noting with Polanyi that, for capital, there have been clusters of characteristic problems with the attempts to commodify labour, money and land. At times economic theory has almost recognized this point.

But we maybe should be particularly careful not to follow Polanyi and Marx in attaching such overwhelming importance to the commodification of labour power. Of course the human effects of the social process we have been anatomizing have been absolutely dreadful at times, and always disturbing at best, but we must tell the story objectively, without privileging any one commodity with disproportionate attention. To our specially morally trained eyes, the use by some humans of others as means to private economic ends condemns itself as soon as it is pointed to, but the point would hardly have occurred to 18th century gentlemen or gentlewomen and the society in which the commodification of life went furthest was one they were apt simply to regard as theirs. I will conclude by pointing out somewhat optimistically that the process they presided over has never yet gone so far as to absorb absolutely all of life. Thereby I hope I have explained why I put the "Very Nearly" in my title. May it always remain nearly capitalist at most!

10.7 AFTERWORD

This essay has a rather specific sounding title but a considerably grander underlying purpose. At its most ambitious level it is intended to initiate a project in historical sociology which could serve to help reunify the historical understanding, and thereby enhance the self-knowledge, of the inhabitants of the globe as they are decanted into the twenty-first century. It takes its starting point from a deceptively simple axiom: modern history is the history above all of capitalism.[30] That bold and yet bland statement contains many subtle implications, even once the complex issues of defining "modern" and "capitalism" have been addressed. For example, many fascinating questions arise as to the nature of human social arrangements prior to the advent of capitalism. Were they inherently more varied previously? And is it only capitalism that has been powerful enough to prevail against all other social forms? If so, how exactly has it been able to do this? Will it last forever? Is it the only form modernity can take? These questions are all both intrinsically interesting and of great importance. None but a fool would deny that the influence of ideas associated with capitalism continues to be immense, and likewise only a fool would deny that capitalism gives rise to a whole variety of special problems, that have to be classified over a wide range from the psycho-social through the socio-political to the physico-ecological. I submit that we cannot know too much about these complex matters and I do not expect to be contradicted on that score! What I also plan to argue however, and which is virtually bound to be controversial, is that we have not yet developed the proper way to set about studying the role of capitalism in "directing" modern history. My guess is that this is part of the reason for the failure thus far of socialist ideologies to present a clearly attractive alternative. The particular aim of this essay then is to introduce and defend a new and, as I see it, more fruitful way of embarking on a much larger project. For a whole complex of reasons which I have expounded elsewhere and which it would be otiose to repeat here, I believe that the question of method is intricately linked to the issue of selecting a case-study.[31] Suffice it to say here that I remain convinced that political economy's original obsession with the English example was and is fully justifiable.

It is my intention eventually to write a book laying out in much more detail the approach sketched here. The book will bear the same main title but the sub-title will be changed to: "Modern English History and the Enactment of Political Economy". Because of this intention to expand the present effort I would particularly welcome helpful comments and suggestions from readers.

Notes and References

1. Uno's original, *Keizai Genron*, was published in 1964. Sekine first introduced Uno's work to the English-speaking world in his "Uno-Riron: a Japanese Contribution to Marxian Political Economy", which appeared in *The Journal of Economic Literature* in 1975.
2. Preliminary English language versions of the two volumes of this work have been published in Japan: Vol. I by Yushindo Press (1984) and Vol. II by Toshindo Press (1986), both of Tokyo.
3. This work was first published in 1944. It was written in ignorance of the work of Uno, who indeed had published relatively little at that point. I have already discussed the general parallelism between the works of Uno and Polanyi elsewhere, in my "Under the Cloud of Capital: History vs. Theory", *Science & Society* (1983), and with Makoto Maruyama in our "The Japanese Counterpart to Karl Polanyi: the Power and Limitations of Kozo Uno's Perspective", published in *The Life and Work of Karl Polanyi*, the volume edited by Kari Polanyi-Levitt of the proceedings of the First International Karl Polanyi conference held in Budapest in 1986.
4. There exists an as yet unfinished translation by Professor Sekine of the 1971 revised edition of a work by Uno entitled *Types of Economic Policies under Capitalism*, in which Uno sketches out an ordered sequence of historical policies.
5. *A Japanese Reconstruction of Marxist Theory* (1986) and A *Japanese Approach to Stages of Capitalist Development* (1991).
6. In my (1983) piece referred to in note 3 above I had already gone so far as to argue that it's not that *Capital* is about a hidden reality, but that it tells us about an ideal which was, as it happens, nearly achieved in England.
7. This passage is from the "Translator's Introduction" to *Principles of Political Economy*, p. xiv.
8. The only serious alternative approach to doing social history comprehensively is exemplified in Fernand Braudel's masterpiece of environmental history, *The Mediterranean and the Mediterranean World in the Age of Philip the Second* (1972, 1973). In that work to a marked degree, and somewhat less successfully in the first two (and only) volumes of his last and lamentably uncompleted work on the history of France, Braudel anchors his discussion of society in a detailed consideration of the peculiarites of place relevant to that society.

9. The use of the familiar Marxist metaphor from the construction trade is deliberate, if knowingly provocative perhaps. I would just remind the over-sensitive that when a Marxist uses some word connoting "foundations" one can nowise simply infer any relative disbelief in the reality of walls and roofs. Indeed it is quite incoherent to think that the main constituent elements of a building can be ranked in importance. Obviously they are all needed, and in equal measure. Which one constructs first however does matter, and in that respect common sense is a perfectly reliable guide!
10. In addition to the now canonical work of E. P. Thompson in this area, *The Making of the English Working Class* (1968), and his only recently published collection of papers, *Customs in Common: Studies in Traditional Popular Culture* (1991), there has been much interesting recent work on the ideologies present in England at the time of the so-called "Industrial Revolution". Boyd Hilton's *The Age of Atonement: the Influence of Evangelicalism on Social and Economic Thought, 1785–1865* (1988), places great emphasis on religion as did Leonore Davidoff and Catherine Hall's *Family Fortunes: Men and Women of the English Middle Class, 1750–1850* (1987), which also discussed the mutual influence of gender stereotypes and organized religion in a way full of implications for political economy. Jeanette Neeson's forthcoming *Commoners* will also have great relevance to the problem of elucidating how capitalism came to settle upon England.
11. It may be interesting to readers to explore an example of the confusion such misfocused critical zeal can lead to. In his masterly *Natives & Newcomers*, (1985) Bruce Trigger chastises the "substantivist" followers of Karl Polanyi (in particular Polanyi's student, Abraham Rotstein) for making too much of the "formal/substantive" distinction and 'forgetting' that acts of "optimizing behaviour" could and did occur even in non-disembedded economies (p. 193). What Trigger seems not to realize is that the analytical point of Polanyi's work was not to say that no "rational" – seeming economic behaviour ever took place prior to capitalism but that under capitalism all other forms of behaviour having a bearing on livelihood are radically undercut. Only one putative motive remains allowable. Previously there had been a set of them, as it were. It is perhaps inevitable that in bourgeois culture the question of motive should bulk so large as to obscure the more puzzling sociological implications of the integrated market system. For Trigger also fails to realize that "optimizing behaviour", to be truly rational, must apply to all activities connected with all goods. In any economy in which only a small proportion of goods are commodified the sphere of application of the calculus may be so limited as to be not really social at all.
12. Karl Polanyi, *The Livelihood of Man* (1977), p. 20.
13. Ernesto Laclau and Chantal Mouffe in their *Hegemony & Socialist Strategy* (1985) attempt to trace Stalinist tendencies to the urge to embrace all of social reality in one theory. That their case is ill-argued, not merely "overstated', is the burden of my "On the Fear of Economism", solicited for *Studies in Political Economy* in 1992.

14. There is much of interest on this point in A.O. Hirschman's *The Passions and the Interests: Political Arguments for Capitalism Before its Triumph* (1977).
15. I believe that one could construct a theory of a purely peasant economy to complement Uno's theory of a purely capitalist society, since it is essential for capitalists that society exists, but a matter of contingent interest to a peasant whether any other households exist.
16. An excellent study which brings out this point very dramatically is Robin Jenkins *The Road to Alto* (1979).
17. I have laid out fairly systematically some of the issues involved in my "On Identifying a Sound Environmental Ethic in History: Prolegomena to any Future Environmental History", published in *Environmental History Review* in 1991.
18. On March 10th, 1801 there was a census done in England. Contemporaries found the degree of urbanization revealed thereby to be surprisingly large but we should marvel more I think at the amazingly even spread of the vast bulk of the population. At that time England had a higher proportion of town-dwellers than any other European country, but it was unusual in having only the one urban area with more than 100,000 inhabitants. Only 19% of the population lived in towns larger than 20,000. Leaving London out of that calculation would reduce that proportion to just over 7%. More than half of the area of England carried a human density of 100 to 200 per square mile. The preceding information was culled from H.C. Prince's contribution to H.C. Darby (ed.), *A New Historical Geography of England After 1600* (1976) pp. 90–4.
19. See the description of the English retail sector in Mui and Mui, *Shops and Shopkeeping in 18th Century England* (1989).
20. F. M. L. Thompson emhasizes this point in his contribution to *The English Landscape* (1985), a collection of lectures edited by S. R. J. Woodell.
21. In my "Legal Protection for the Soil of England: the Spurious Context of Nineteenth Century 'Progress'", published in *Agricultural History* in 1992, I discussed several other aspects of nineteenth century developments which would have disturbed and disappointed Smith.
22. As William Cronon has shown in a masterful way for the case of Chicago in his *Nature's Metropolis* (1991), some cities were built up purely speculatively, as it were, and then went on to completely shape huge hinterlands for their own convenience. In England until about 1850 something like the opposite dynamic obtained and the state of the countryside dictated much about patterns of urban development. See the section on "Open Field Towns" in Chapter 9 of W. G. Hoskins, *The Making of the English Countryside* (1985).
23. Avner Offer in his *The First World War: An Agrarian Interpretation* (1989) makes a strong case for the view that the relative prosperity in farming before the war was also fundamentally military in origin.
24. See his *The Overdeveloped Nations*.
25. See John Bossy's, *Christianity in the West: 1400–1700* (1985) for a very sophisticated anthropological account of the content and sociological

implications of theological disputation. This work, though short, is much more complex in theoretical terms than the pioneering sketches done by Weber and Tawney.

26. It is striking in this context to see how deeply ahistorical even conservative contemporary politicians are. Thatcher expressly sought to restore what she called "Victorian values" conveniently forgetting the profound role of evengelical religion in the supposedly aggressively greedy lifestyles she sought to bring back. Judging by what happened in England when most capitalists were evangelical Christians, the prospect of today's Tories reviving the capitalist ethos minus the Christianity is disturbing indeed.
27. The excellent recent biography by Frances Harris of Sarah, Duchess of Marlborough, called, very aptly, *A Passion for Government* (1992), shows how one of the most important non-royal personages of the period after 1688 felt she had constantly to be on the watch for pernicious political developments right up until her death in 1744. The sustained constitutional tension exhibited in this life is altogether remarkable.
28. See my (1992) article referred to above for several relevant quotations from Smith.
29. The sample budgets gathered in John Burnett's *Plenty and Want* (1989) make alarming reading but from the point of view of the research project being described here their sheer simplicity is extremely heartening.
30. I owe the insight that historians can be said to work with axioms to my colleague, Professor P. K. Christianson.
31. See my forthcoming book, *The Centrality of Agriculture.*

11 Regulation Theory: A Critique

Robert Albritton

Regulation Theory has proven to be an immensely attractive research program. From its beginnings with the publication of Aglietta's *A Theory of Capitalist Regulation* (1979) in the mid-1970s, it has continued to attract many adherents through the 1980s to become probably the largest and most influential school of Marxian political economy. Part of its attractiveness is due to its being a highly flexible research program able to absorb criticisms and develop variations. Indeed, the variations are sufficiently different and numerous to support viewing Regulation Theory not as a single approach but as a cluster of approaches unified by "family resemblances" but no core. In a recent article Jessop (1990, 154) attempts not only to counteract the centripetal tendencies of Regulation Theory, but also attempts to counter adaptations of it that would neglect class struggle in favour of concern with "structural cohesion". In order to carry out this project, Jessop claims that it is important to return to the pioneering texts. I shall also return to the pioneering texts, particularly Aglietta's (1979) *A Theory of Capitalist Regulation*, but my orientation and purpose will be quite different from Jessop's. My aim is to point to weaknesses and to options missed in Aglietta's pioneering text in order to support an approach that can be viewed either as a substantial reform of regulation theory or as an alternative approach with some important similarities.

Before launching my critique, I want to underline the important contributions made by Regulation Theory. These contributions fall into five main areas. First, by attempting to construct an intermediate level of theory connecting abstract economic theory and history, they focus our attention on the importance of levels of analysis in Marxian political economy and particularly on the need for a mid-range theory. Second, by attempting to consider non-economic social variables in connection with economic variables, they have produced a more institutional economic theory. Third, their efforts to conceptualize distinct regimes of accumulation, focuses attention on the problem of periodizing capitalist history. Fourth, their particular formulation of mid-range theory has focused attention on the importance of the labour process. Fifth, their conceptualization of "fordism" has focused attention and debate on the theorization of post-World War II capitalism. In this paper, I am mainly interested in how Aglietta, in his founding work, theorized capitalism at the level of abstract economic theory and how he theorized different levels of abstraction and their interrelations.

Aglietta (1979, 9) starts his book by pointing to the growing criticism of the dominant economic theory (general equilibrium theory), which fundamentally fails to give any adequate account of the historical evolution of economic phenomena or of how the economic relates to the social. I agree with Aglietta and would go further and claim that any general economic theory should be assessed above all in accord with how effectively it relates abstract economic theory with concrete historical change and how effectively it relates the economic with the social. But I also believe that how abstract economic theory itself is conceptualized may be crucial in how effectively the abstract is related to the concrete and how the economic is related to the social. Hence, the attention paid in this paper to how Aglietta formulated value theory, and the implications of his formulations for the above two issues.

My approach to Marxian political economy is fundamentally informed by the work of Japanese political economists Uno and Sekine.[1] The approach that I have developed based on their work will be used to question and criticize Aglietta's interpretation of Marx's *Capital*, his use of levels of analysis that mediate between abstract and concrete, and his understanding of how the economic relates to the social.

11.1 PHILOSOPHICAL ECONOMICS

"Philosophical economics" may sound like a contradiction in terms since most economic theory is unphilosophical. Unlike most other economic theorists, however, Aglietta does attempt to outline the philosophical underpinnings of his theory. In my view, a philosophical economics ought to concern itself first and foremost with the peculiarities of the economic as an object of knowledge. In particular we need to understand the basis for making the abstractions that constitute economic theory, and, as a corollary, how the abstract economic relates to the historical and social. Aglietta confronts these issues primarily by criticizing classical general equilibrium theory. For example, he argues (1979, 11) that equilibrium theory cannot adequately understand the historical because it starts with "a conception of time that renders dynamics a mere variant of statics – in effect, a logical time which is not the expression of any real movement". Such an approach cannot come to grips with the reproduction and accompanying ruptures of real socio-economic systems, but instead turns economic growth into something that is automatic, linear, abstract, and finally ahistorical. General equilibrium theory also cannot adequately understand the social because it starts with the assumption of a "rational and sovereign subject, free of any social tie . . . The goal of theory is to express the essence of its object by stripping it of everything contingent; institutions, social interactions, conflicts, are so much dross to be purged to rediscover economic behaviour in its pure state" (1979, 14).

For the concept "equilibrium", Aglietta wants to substitute the concept "reproduction", because it forces us to break with the automaticity of "equilibrium" and to think concretely about the socio-historical conditions of that which exists. "To speak of reproduction is to show the processes which permit what exists to go on existing" (1979, 12). We need "a theory of the regulation of capitalism which isolates the conditions, rhythms, and forms of its social transformations" (1979, 15). Such a theory would analyze the economic system as a whole in order to delineate "the general laws that are socially determinate" and their "historical conditions of validity" (1979, 15). The general laws of regulation theory are to be sharply distinguished from those of general equilibrium theory. As Aglietta puts it: "The study of capitalist regulation, therefore, cannot be the investigation of abstract economic laws. It is the study of the transformation of social relations as it creates new forms that are both economic and non-economic, that are organized in structures and themselves reproduce a determinant structure, the mode of production" (1979, 16). It would offer a "complete alternative to general equilibrium theory" (1979, 13).

Aglietta claims (1979, 380) that his regulation theory is based on "the general theory of capitalism founded by Marx", or, in other words Marx's *Capital*. And yet he does not systematically discuss his interpretation of *Capital* in order to distinguish it from other interpretations or to show where he differs from Marx. It is entirely possible, therefore, that regulation theory is not based so much on Marx's *Capital* as on Aglietta's free-wheeling extraction and refashioning of Marx's concepts. The concern here is not that one cannot revise Marx, but rather that such revision should be up front and defended.

By the way Aglietta frames his argument, our choice seems to be between general equilibrium theory and regulation theory, or between making "equilibrium" our central concept and making "reproduction" our central concept. But, as I hope to show, there are other alternatives–alternatives that are closer to Marx's original formulations and that can potentially more effectively relate the economic with the historical and social.

Marx used both "equilibrium" and "reproduction" in *Capital*, but did not use "equilibrium" as does classical general equilibrium theory and did not use "reproduction" as does Aglietta. Marx argues that there are tendencies towards equilibrium in capitalism and that capital's immanent laws could not be theorized were there no such tendencies, but he also argues that periodic crises disrupt these tendencies so that equilibrium is never reached as the equilibrating forces are continually disrupted.

Marx uses "reproduction" primarily in connection with the "reproduction schema", where the aim is not to consider the concrete reproduction of a social formation, but to show the abstract possibility of a balanced circular flow between the production of capital goods and consumer goods in the expanded reproduction of capital. Perhaps Marx anticipated the traps of crude structural-functionalism that might stem from placing the concept "reproduction" at the center of economic theory. This is because in concrete reality, social practices are reproduced in the most diverse ways over the most diverse spatialities from the local to the global and over the most diverse time spans from seconds to centuries. A great deal of social reproduction is not the same as economic reproduction. Furthermore, "reproduction" may not be the best concept to theorize the continual change and renegotiation of social practices that results from power and resistance to it, since often social practices are not reproduced but are transformed. And the tendency with "reproduction" is to focus on the reproduction of some limited set of economic practices that becomes the "invariant structure" with the social totality evolving as a function of that structure. For Aglietta, the "invariant kernel" is the capital/labour relation; the general laws of its reproduction explain the regulation of the capitalist mode of production, and the capitalist mode of production is the determinant structure that explains the evolution of the social totality.

Aglietta equates "pure economic theory" with general equilibrium theory, and as a result fails to understand how Marx's *Capital* can be read as a "pure" economic theory, but in a sense that is different from classical general equilibrium theory. Where general equilibrium theory starts with the abstraction "rational economic man", Marx starts with a totally commodified economy in which capital as "self-expanding value" can expand itself without reliance upon extra-economic force precisely because all inputs and outputs of the capitalistic production of commodities are totally commodified and hence market regulated. Marx's theoretical object is based on the fact that capital is self-reifying, or, in other words, that social relations tend to be reduced to a cash nexus by the expansion and deepening of a commodity-economic logic inherent to capital's expanded reproduction. Unlike general equilibrium theory, Marx does not express the essence of his object "by stripping it of everything contingent", instead he lets the self-reifying tendencies of capital complete themselves in theory to the point where persons become mere bearers of economic categories. The result is not a universal theory of equilibrium based on voluntaristic abstractions, but a theory of capital's inner logic based on historically specific reified social relations. Thus the contingent falls away not because of a voluntaristic act of "stripping", but because in a society consisting of social relations between things and material relations between persons, the social becomes totally absorbed into the economic. In conclusion we can say that Marx's theory of capital's immanent laws is a "pure" economic theory, achieved not by a theoretical voluntarism that "purges" the social, but by extending to completion capital's self-reifying tendencies that absorb the social into the economic at the point of total reification.

Understanding the self-reifying character of capitalism is absolutely crucial to understanding its uniqueness as a theoretical object.[2] No other object of social theory is self-reifying in the same sense and to the same degree. This special characteristic of capital explains why it is possible to theorize an inner logic or inner essence without voluntaristic abstraction. But if the social is absorbed into the economic only in the context of pure capitalism, then as we move from this most abstract theory to more concrete levels of analysis, the social must reemerge from the economic.[3] In this way we can see how a "pure" economic theory in the case of capital becomes possible without abstraction from the social because the reifying force of the economic simply absorbs the social. Furthermore, since at the level of pure economic theory persons are mere "personifications of economic categories", and at the level of historical analysis persons may collectively impact on the course of history, it is apparent that at least two distinct levels of theory are necessary: one where agency is absorbed into structure and one where it is not.

I would claim that one of the weaknesses of Marx's *Capital* is his failure to adequately problematize the relation between logical time and historical time.[4] If historical time is simply a function of logical time as Aglietta (1979, 11) claims it is with general equilibrium theory, then it is true that dynamics would be a mere function of statics. Surely this is to be avoided in economic theory. On the other hand, to collapse logical time and historical time together as Aglietta tends to do must either undermine the rigour of economic theory, produce an economic reductionist history, or both. As with the relation between the economic and the social, I would argue the solution is separate theories for distinct levels of analysis. The theory of capital's immanent laws is primarily logical time. For example, the tendency for capital to centralize is at this level an abstract logical tendency. We cannot determine whether or not centralization would lead to monopoly in 5 years, 50 years, 500 years, or never, because at the level of concrete historical time all sorts of factors may retard, block, or even reverse this tendency. It is important to have a theory of capital's abstract tendencies in logical time in order to fully understand its character and operating principles, but it is also important not to make historical time a function of logical time, and for that we need distinct theories operating at the level of historical time.

Aglietta claims that Regulation Theory must analyze the "economic system as a whole" (1979, 15). But he does not carefully analyze the implications of locating the whole as he does within a particular national society. Why should not the "economic system as a whole" refer to the global system, and since, it apparently does not, how does a particular country's economy relate to the global system? Furthermore, Aglietta is not clear in his use of "mode of production". What precisely does this crucial concept mean? How does the theory of the capitalist mode of production relate to international society and to particular national societies? According to Aglietta "To speak of the regulation of a mode of production is to try to formulate in general laws the way in which the determinant structure of *a society* is reproduced" (1979, 12, my emphasis). How do the general laws governing the reproduction of the determinant structure in one society relate to those of another society, and how do such different systems of general law relate to the theory of the capitalist mode of production? Aglietta's lack of clarity on these issues has produced confusions and tensions in the subsequent development of regulation theory on both the question of levels of abstraction in economic theory and on the question of relating global, regional, national, and local economic configurations. His use of terms like "mode of production", "economic system", "social system" raise more questions than they answer.

To summarize the argument so far, Aglietta begins by mapping out Regulation Theory in opposition to general equilibrium theory. While I would agree in general with his critique of classical general equilibrium theory, the alternative that he presents is not the only possible alternative and it is problematic. In particular, he fails to grasp the uniqueness of capital as a theoretical object, a uniqueness that is absolutely crucial to Marx's theoretical project. As a result, he cannot understand how Marx could formulate a theory of abstract economic laws without voluntaristically stripping away the historical and social. Aglietta's solution is to create a mid-range economic theory that while integrating the social and historical with the economic to some extent, still bears strong economistic tendencies. It is a theory that borrows categories from Marx's *Capital* quite selectively and employs them quite differently from Marx. It is a theory that has a certain richness and power precisely because it attempts to integrate the economic with the social and historical, but given its epistemological framework, it cannot avoid strong tendencies towards economic reductionism. Aglietta rejects Marx's immanent laws of capital, but borrows the idea that the exploitation of labour is central. Based on this idea, Aglietta attempts to create his own formal model of value categories, which is not only based on voluntaristic abstraction, but is really quite otiose to his mid-range theory. The empty formalism of Aglietta's value theory is evidenced by its subsequent abandonment even by Aglietta himself.

11.2 VALUE THEORY

In general Aglietta presents his value theory as if it had no substantial differences from Marx's basic theory of capital.[5] One difference, he does note in passing is his rejection of Marx's idea that labour-power is a commodity. Since this idea is absolutely crucial to Marx's entire theory, it is peculiar that Aglietta dismisses it by simply labelling it a "classical fiction" (1979, 31). Marx makes it very clear that labour-power is not a commodity like other commodities, and yet it is a commodity in the sense that a capitalist must go to the labour market to purchase labour-power as a commodity input in a way that is similar to the purchase of other material commodity inputs. Marx's fundamental formula for capital, M – C – M′ – the use of money to make more money through the purchase and sale of commodities – would make no sense without the commodification of labour-power. It follows that the commodification of labour-power is an absolutely crucial and fundamental category to his entire theory.

For Marx, capital is fundamentally the subsumption of the labour and production process to the commodity form, to a commodity economic logic, or in other words to "self-expanding value". Furthermore, it is the commodification of labour-power that enables abstract labour to operate as the substance of value. In contrast, Aglietta ignores the importance of Marx's theory of circulation forms and proceeds directly to "the *invariant kernel* of the capitalist mode of production, defining it as the mode of organization of social labour" (1979, 37, my emphasis). The starting point for Aglietta's value theory is surplus value because "it defines the economic form in which the labour of society is appropriated" (1979, 37). But in order to understand surplus-value, it is first necessary to understand value, which Aglietta defines as expressing "the relations by which particular labour performed in different sites where productive forces are gathered together becomes social labour" (1979, 37).

How can we understand the processes by which abstract labour becomes the substance of value without a theory of the circulation forms commodity, money, and capital? Aglietta solves this problem with a theoretical *deus ex machina*, the "homogeneous space of value". Value categories, including "abstract labour", become abstract because they participate in the homogeneous space of value. Aglietta's (1979, 38) claim that "In economics, the task of abstraction is possible because a process of homogenization exists in the reality to be studied" strikes me as true. But Aglietta does not theoretically lay out this process of homogenization other than to say that it is produced by exchange (1979, 53). In my view the process of homogenization is a process of reification which only becomes complete when the self-organization of persons is entirely subsumed to the self-organization of capital.

The consequences of not clearly separating levels of analysis can lead to economic reductionism and to reproducing in thought a "market mentality".[6] In a purely capitalist society we assume that all labour produces commodities for capitalists, but in reality this is never the case. At all times in the history of capitalism a huge amount of labour is not capitalist wage-labour and hence is not valued by capital.[7] The most obvious example would be domestic labour. It is important to separate the theory of a purely capitalist society, which ignores domestic labour because it does not fall within the purview of capital as self-expanding value, and more concrete levels of analysis where we need to explore the interrelations between capitalist wage labour and other kinds of labour. Failure to make this separation reproduces the way capital values things at concrete levels of analysis where we should be challenging and resisting such valuation. The total labour of society is abstract labour only in the context of pure capitalism where the purview is that of capital, which limits itself to those elements that fall within self-expanding value. But as Aglietta moves towards the concrete, abstract labour and the wage-relation remain the invariant core and in a sense the prime mover of capitalist history. Such a conceptual framework will tend to marginalize any labour not fully constituted within the wage-relation and the actual complexity of types of labour and their interrelations cannot be effectively thought. As we move from the abstract to the concrete, it is important for our categories not to be overly reified so that we can adequately explore the way capital interacts with all types of labour, not just labour that creates surplus value for capital. Aglietta's tendency to equate abstract labour and value with *social* labour must either recognize the social in this context to be the social of pure capitalism, or else his concept of social is one-sidedly exclusionary.

For Aglietta capital is not only equated to the wage relation, but also the wage relation is made out to be the "invariant kernel" of the capitalist mode of production, out of which other value categories are generated. Thus, for example, surplus value is generated because "the wage relation *effects a division* of the general space of value by dividing total abstract labour (VA) into value of labour-power (V) and surplus-value (SV)" (1979, 46, my emphasis). Instead of Marx's very clear analysis of surplus value as a difference between total value created by labour power and the value of that labour power, we get the entirely mystical idea that the wage relation "effects a division".

After formally expressing a number of relationships between constant capital, variable capital, and surplus value, Aglietta declares that "All these relationships are defined in the homogeneous space of value, and have no meaning outside of this space" (1979, 54). If this is the case, it is hard to see how his various value categories that ground his laws of capitalist accumulation can be used in connection with more concrete levels of analysis which always involve non-homogeneous space. Nonetheless, there is no doubt that on this point he violates his own strictures and does apply value categories at more concrete levels of analysis. He makes the very problematic move from homogeneous to non-homogeneous space without developing the complicated mediations and transitions involved. How exactly does the homogeneous space of value relate to the non homogeneous space of capitalist history?

I have claimed above that abstract labour only becomes clearly and fully the substance of value in a purely capitalist society because only there are social relations totally reified. Further, only there does all private labour become a fraction of total social labour, because only there is all production the capitalist production of commodities by commodified labour-power. But a purely capitalist society never exists in history, and the ways and degrees in which labour power is commodified vary from stage to stage of capitalist development. For example, take the putting-out system that was so widespread in England in the 18th century. Typically cottage workers would be provided with inputs and would be paid a piece wage for their output. In this case it is difficult to see how the wage relation "effects a division" that generates surplus value. Aglietta's capital-as-wage-relation cannot adequately theorize the situation where merchant capital indirectly exploits wage labour through a putting-out system. Presumably by "wage relation" Aglietta means to refer to the capital/labour relation, one form of which under capitalism *is* the wage relation. The problem is that capital is more than the capital/labour relation, and the wage relation is only part of the capital/labour relation. Furthermore, the wage relation itself varies qualitatively depending on the ways and degrees to which the labour and production process is subsumed to M – C – M′. For example, the wage relation is radically different when the labour and production process is only formally subsumed (as in the putting-out system) as opposed to substantially subsumed (as in factory production).

One example of the problematic character of Aglietta's move from the homogeneous to the non-homogeneous is his use of the categories absolute and relative surplus-value. In the context of a purely capitalist society absolute surplus value is logically prior because it is based on the length of the working day, which is the basic reference unit for surplus value in general. Surplus value is always absolute because it is based on the working day, and absolute surplus value can be expanded by lengthening the working day. Surplus value is also always relative because the working day is always divided into the time needed to create the value of labour power and the time in excess of this during which surplus value is created. Relative surplus value is increased by productivity increases that cheapen the value of the commodities that constitute the value of labour power. Since the length of the working day is the framework within which relative surplus is generated, all surplus value is both absolute and relative. Aglietta (1979,51) himself realizes this when he writes: "Absolute and relative surplus value are thus indissociable from one another." Given this "indissociability" and given that the space of value is homogeneous, one must be careful in arguing that one phase of capitalist history is characterized by absolute surplus value extraction and another by relative. But when Aglietta moves to the more concrete level of analysis associated with his "regimes of accumulation", it becomes apparent that absolute and relative surplus value can not only be used out of the context of homogeneous value space, but also they can be considered at least relatively autonomous since absolute surplus value is the foundation for extensive accumulation and relative surplus value for intensive accumulation.

To apply these two categories to historical analysis and claim that in one stage of history absolute surplus value predominates and at another relative surplus value predominates is a common folly. In the first place, the category "surplus value" assumes fully homogenized abstract labour, where the labour market is not interfered with by the state, by capital, or by labour. In the second place, even if one wants to differentiate between stages of capitalist development according to whether more emphasis is placed on lengthening the working day or on increasing productivity, then one should use this language and not the concepts "absolute and relative surplus value" that even Aglietta admits belong to the realm of the "homogeneous space of value" and "have no meaning outside this space".

I want to go further and argue that periodizing capitalist development in this manner is not accurate. For example, was lengthening the working day or increasing productivity more important to the putting-out system in England in the 18th century? Since the cottage workers controlled their own labour process, merchant capital had no control over the length of the working day and little control over productivity either. In fact since putting-out work was sporadic, the length of the working day averaged over a year would have been relatively short. Productivity, on the other hand, increased considerably both because of small but important changes in the division of labour and because of improved tools. Improvements in the spinning wheel and the hand loom more than doubled productivity in woollen manufacturing in the first half of the 18th century (Heaton, 1965, 310). It would seem then, that increased productivity was more important than lengthening the working day for increasing profits even in this early stage of capitalist development.

What about cotton manufacturing, the center of the capitalist factory system of the 19th century? While the working day was extended to some extent until limited by legislation, it is not clear that this had a big impact on profits since it is likely that with the longer working day the pace of work was less intense. In contrast, productivity increased by huge amounts, it being estimated that the perfected self-acting mule increased spinning productivity over the old-fashioned standard spinning wheel by as much as 300 fold (Crouzet, 1982, 101). What I am suggesting is that at the historical level of analysis productivity increases are always more important to capitalist profits than lengthening the working day, and that this distinction therefore provides no clear basis for periodizing capitalist development.[8]

Aglietta's appropriation of Marx's "reproduction schema" is also highly problematic. Having generated surplus value as an *effect* of the wage relation, Aglietta then goes on to introduce Marx's reproduction schema directly into his elucidation of relative surplus value. Aglietta simply asserts that "the motive impulses in the transformation of the forces of production, in effect, derive from Department I" (1979, 56). He concludes from this that the two Departments, will tend to develop unevenly, and later on, this totally undemonstrated assertion comes to play an important role in his analysis of history as a movement from extensive to intensive regimes of accumulation and in his analysis of capitalist crises.

Marx uses his reproduction schema at the end of Volume II for a very limited and specific purpose. He wants to indicate in a circular flow model the abstract possibility of an expanded reproduction of capital.[9] He very specifically excludes any consideration of changes in value relations of the sort involved, say, in a rising organic composition of capital. I suspect that one reason that he did this was precisely to avoid developing the Reproduction Schema into anything like an equilibrium theory, where crises could be understood in some simplistic and mechanical way as being generated out of an disequilibrium between Departments I and II. But Aglietta does use the Reproduction Schema in this mechanical way, which is surprising given his aversion to "general equilibrium theory".

Aglietta's "laws of competition" might seem to correspond roughly to Marx's theory of distribution in Volume III of *Capital*, because they both deal with inter-capitalist relations. This, however, is practically their only similarity. Instead of considering the relation of capital to land and inter-capitalist relations in the context of value categories, Aglietta's laws of competition descend rapidly towards the concrete where they consider historical trends towards monopoly, state intervention, inflation, and crisis.

In order to develop his theory of price determination, Aglietta reverts back to his equilibrium interpretation of Marx's Reproduction Schema. Aglietta's "reproduction" functions as a trojan horse for bringing "equilibrium" back into the center of marxian political economy. He argues that a general rate of profit only exists to the extent that the two Departments are in equilibrium (1979, 285). As discussed above, Marx does not use the Reproduction Schema in any immediate sense in his theory of price determination, which centers on the transformation of values into prices, nor does he generate his theory of crises directly out of those same Schema as Aglietta does. Aglietta goes on to claim that his theory of production prices "directly permits a concrete analysis, because it is linked to a conceptualization of historical time" (1979, 295). I find this claim extremely revealing of the strong strains of economism in Aglietta's theory where abstract economic theory is imposed directly on historical evolution.

According to Marx, Department I produces means of production and Department II produces means of consumption, requiring that an exchange take place between them. Marx uses these Schema to show in value terms the possibility of the expanded reproduction of an economic system that is basically anarchical. The Reproduction Schema as value categories belong to what Aglietta has called the homogeneous space of value. But, as I have argued previously, Aglietta is not at all clear about what is involved in making this space homogeneous – it appears to be the automatic creation of exchange equivalences. Unlike Marx, Aglietta introduces changes in the organic composition of capital into the Schema, and since such changes imply changes in the forces of production, he seems to assume that this automatically links the Reproduction Schema to historical time. I see no reason whatsoever to accept this assumption since changes in the forces of production can just as easily be conceptualized in logical time. The crux of the problem is that Aglietta apparently sees little difficulty in moving from abstract to concrete or from homogeneous to non-homogeneous. As a result there are strong tendencies for economism to emerge, in which an homogenized abstract is imposed upon a non-homogenized concrete.

According to Aglietta "the laws of competition derive rigorously from the law of accumulation. The sum total of these laws *constitutes the metabolism of the social formations subjected to the capitalist mode of production*" (1979, 273, my emphasis). Here he moves directly from mode of production to social formations suggesting that his laws are their inner essence. But the articulation of modes of production in social formations is a notoriously complex problem. Social formations are very diverse, and they are subjected to the capitalist mode of production in different ways and to varying extents.

In Aglietta's theory the laws of competition are grounded upon the abstract tendencies for capital to concentrate and centralize. But as abstract tendencies these tendencies have no particular historical time attached to them – the move from competitive to monopoly capitalism would not necessarily ever occur in the context of pure capitalism because concentration would lead to larger but not fewer firms and substantial centralization would not necessarily occur because in pure capitalism we cannot assume the existence of the limited liability joint stock company. Even if there were some tendency towards monopoly in pure capitalism, as pointed out earlier, it could take five years or five hundred years. For example, in order to understand, the incredibly rapid merger movement in the United States between 1898 and 1902, all sorts of historical contingencies and specificities need to be taken into account.[10] And yet, Aglietta proceeds as if such centralization is a necessary outcome of the workings of his inner law of accumulation. Indeed, monopolistic corporations are not only the outcome of the working of his laws, but also he sees no difficulties posed for his homogeneous space of value by the existence of monopoly.

Aglietta's laws of competition are just as highly selective and idiosyncratic as are his laws of accumulation. For example, he does not systematically consider the distribution of surplus value between industrial capital, commercial capital, interest-bearing capital, and landlords as Marx does. I believe that to make sense of these relationships, it is necessary to assume initially a purely capitalist society where value relations generally hold. Instead Aglietta uses his laws of competition to rapidly descend to more concrete considerations such as financial centralization and post-world war II inflation. As a result, his iron laws extend quite far towards the concrete where one would like to see more space for particularity, contingency, and agency.

Summarizing the argument so far, it appears that Aglietta's interpretation of Marx's *Capital* is highly selective and one-sided. I do not for a moment believe that one must remain absolutely true to Marx's original. Aglietta, however, presents himself as simply elaborating Marx's theory, when in fact he is altering it in all sorts of ways. It is therefore necessary to think through his alterations to see if they are improvements on the original. I have indicated a few ways in which they are not. But I want to qualify this with recognition of an important achievement on Aglietta's part. One-sidedness distorts, but it also may serve to bring something into sharp relief and hence contribute to our understanding. Aglietta's one-sided equation of capital with the wage-relation brings into sharp relief important interconnections between the labour process and the conditions of existence of the working class.

11.3 MID-RANGE THEORY

It is generally agreed that Regulation Theory should be primarily viewed as a mid-range theory. For example, Davis claims that Aglietta's central concern is the "absence of a theoretical level linking class struggles to their structural determinants in the accumulation process" (Davis, 1978, 208). Referring to both the Regulation School and the Capital-Logic School that he has tried to synthesize, Jessop writes: "Both . . . have produced concepts at a middle range, institutional level; both are concerned with stages and phases of capitalist development rather than with the abstract laws of motion; both are sensitive to the relative autonomy of the economic and political spheres. . . ." (Jessop, 1988, 162). Both Davis and Jessop seem attracted by a middle range or a linking theoretical level, and it is true that such a middle range theory could, at least in principle, avoid the economic reductionism the follows from imposing the laws of motion of capital directly on history.

It is clear that Aglietta is not interested in constructing a theory of capital's inner logic. Instead he posits a formal axiomatic model of value categories only as required to inform his mid-level theory of regulation. As Aglietta puts it "By formalizing the law of capital accumulation . . . we introduced the concepts needed for an overall perspective on its historical movement" (1979, 65). It is clear that the homogeneous space of value constructed by Aglietta is not a theory of capital's immanent laws but is a Weberian ideal-type. Furthermore, like Weber, Aglietta freely borrows from abstract economic theory in order to construct a mid-range theory, only he borrows from *Capital* instead of from marginal utility theory as does Weber. For Aglietta, value theory is simply a geometry used to establish certain formal preliminaries to a theory that is essentially mid-range. There is minimal effort to think systematically about the theory of capital's inner logic and its connection to mid-range theory or to think systematically about the relations between mid-range theory and historical analysis as distinct levels of analysis. As a result there tends to be only one level of theory that has any substance: mid-range theory, which tends to absorb the other two levels into a single "spiral" (Jessop, 1990, 189). Aglietta (1979, 47) emphasizes that "It can never be repeated often enough that not all economic problems can be treated at the same level of abstraction", and yet he reduces the importance of the difference between levels of abstraction by treating them all within one theory rather than having distinct levels of theory to deal with different levels of abstraction.

The tendency of Regulation Theory is clearly to theorize the economic first as a regime of accumulation and then add on political and sometimes ideological supports. This tendency can be seen clearly in Boyer's (1990, 11) claim that Regulation Theory is concerned with the "impact of a set of social relations on the stability of the economy". The tendency of his thinking here is clear: an "inside" ("the stability of the economy") is disturbed by an "outside" ("a set of social relations"). In contrast, I would argue that the task of mid-range theory is not in the first instance to indicate how the political and ideological impact on the stability of the economy, but rather to explore the ways in which the political and ideological are implicated in the very constitution of the economic. And to get this clear it is an enormous aid to have a high-range theory, a theory of capital's inner logic to help clarify the basic character of the economic under capitalism. If we start with a clear demarcation of the economic, our ability to analyze its interconnections with the political and ideological is greatly enhanced. Thus, for example, at the level of mid-range theory it would be very important to clarify both the degree to which labour-power is typically commodified at different stages of capitalist development and how politics and ideology are directly implicated in the constitution of the prevailing degree and type of commodification.

In order to do this well, we need to understand the conditions of existence of totally commodified labour-power, and this can only be rigourously clarified in a theory of pure capitalism.

Boyer (1990, 26) claims that Regulation Theorists disagree over the most abstract economic concepts, but agree on the core intermediate notions. And the reason why this is possible is that the theory of value apparently has little bearing on mid-range theory. I suppose that if mid-range theory is to revolve entirely around the organization of the capitalist labour process and the associated exploitation without any concern with questions of degree of commodification or of the precise capitalist character of basic economic categories, then high-range theory could be dispensed with. Boyer (1990, 31, my emphasis) states that the aim of the regulation approach is to use "long-term historical data to enrich and critically elaborate Marxist *intuitions* concerning the dynamics of capitalist economies". And there is little doubt that once value theory is abandoned, as it has been by most Regulation Theorists, Regulation Theory's connection to Marxism will be reduced to little more than "intuitions", intuitions that can be abandoned as easily as they are adopted. I am suggesting that a theory based on "Marxist intuitions" can easily become a theory based on non-Marxist intuitions.

Like Aglietta, Boyer has at least some awareness of the importance of levels of analysis. Boyer (1990, 31) writes: "The problem is to construct different sorts of concepts . . . from the highest level of abstraction to immediate experience." Despite such promising statements, Regulation Theory has made little effort to do this in any sort of rigourous way. It has all but abandoned value theory, and it generally ignores the immense problems of applying mid-range theory directly to the analysis of history. It is my claim that "different sorts of concepts" require different sorts of theories, and that effective mid-range theory must clearly distinguish its concepts and its aims from both "high-theory" and "low-theory".

Boyer (1990, vii) claims that Regulation Theorists "reject the idea of general, eternal laws applicable to all socioeconomic systems". But the immanent laws of Marx's *Capital* do not need to be used this way. With the concept of "the theory of a purely capitalist society" we preserve the rigour of the theory of capital's inner logic, while problematizing its use in informing more concrete levels of analysis. By not seeing this option, Regulation Theory instead produces a mid-range structural-functional theory which is informed in ad hoc ways by "Marxist intuitions". They give up the extremely important informing function that can be played by a carefully developed theory of pure capitalism.

Regulation Theorists continually refer to the importance of levels of abstraction, but never in fact take this idea very seriously. Jessop (1990, 176), for example, claims that it is "the task of regulation theories to extend Marx's analysis from the internal laws of the capitalist mode of production to the contingencies of capitalist accumulation. This task is conducted at different levels of abstraction and helps to realize Marx's own stated objective of analyzing the real concrete as a concrete synthesis of multiple determinations."[11] How are these different levels of abstraction to be related to one another? Jessop (1990, 189) claims that there is a "spiral" from the abstract/simple to the concrete/complex, and that there cannot be a "radical break in the spiral movement of analysis." This does not help much. Can there be breaks that are not radical? How can we differentiate levels of abstraction if they all flow into one another in a smooth spiral? How, in particular, do we theorize the highest levels of abstraction? Marx thought that there was a single inner logic of capital, and that is exactly what he theorized in *Capital.* Jessop's realist interpretation of Regulation Theory must rest upon a clear conceptualization of a theory of capital's inner mechanisms, and yet, Jessop (1990, 189) argues that there is no single logic of capital, but instead "a series of logics with a family resemblance". But this is a radical break with Marx, and requires a whole new thinking of precisely how these multiple logics are to be theorized and related to one another and to Marx's *Capital.* Jessop's only reason for rejecting a single theory of capital's inner logic is his refusal to accept a level of abstraction at which a commodity-economic logic determines economic outcomes. Of course, it is true in concrete reality that the law of value always interacts with "specific modes of regulation and accumulation strategies" and on the "contradictory articulation of commodity and non-commodity forms" (Jessop, 1990, 187–8). Marx was completely aware of this, but it did not stop him from theorizing a single logic of capital in the abstract, in which he assumed a total reification where persons were reduced to mere bearers of economic categories. Jessop seems to want it both ways. On the one hand, he rejects Marx's theory of a single inner logic of capital. On the other hand, he wants some kind of value theory to define "the basic parameters of capitalism" (1990, 187). On the one hand, he wants different levels of abstraction, and even claims that "levels refers to the ontological depth of the real world" (1990, 207). At the same time, he wants an unbroken spiral from abstract to concrete.

11.4 HISTORICAL ANALYSIS

Just as the relation between Marx's *Capital* and mid-range theory is underspecified and confused in Regulation Theory, so is the relation between mid-range theory and low-range theory (historical analysis). This latter confusion produces the continual and unresolved tension between structural-functionalism at the level of mid-range theory and the aim to be sensitive to contingency, agency and complexity at the level of historical analysis. Jessop (1990, 196) warns us against "reintroducing the concept of reproduction and reducing concrete subjects to the necessary *Traeger* (supports) of the dominant structure". Part of his reason for returning to the pioneers of Regulation Theory is that they emphasized class struggle as opposed to "recent regulationist studies" which "have often focused on questions of structural cohesion and neglected social agency" (1990, 154). But, as I have argued, it is precisely the pioneers, Aglietta in particular, who have placed "reproduction" at the center of regulation theory. According to Aglietta, the state is "a mode of social cohesion required by relations of production" (1979, 26). In other words, the state is a function of the economic, or, to be more precise, it is a function of the necessity to maintain the wage-relation. Aglietta's claim that "state monopoly capitalism is the mode of articulation of the structural forms engendered by Fordism", avoids functionalist language, but in the end says little more than that state monopoly capitalism is a function of Fordism (1979, 383). Finally, Aglietta (1979, 382) claims that "the field of capitalist social relations is . . . a complex whole structured by the domination of the wage relation". Despite the Althusserian language here that brings in "complex", the fact that society is a whole or a unity structured by the wage relation implies not that the whole is a simple function of the wage relation, but at least implies that it is the wage relation that structures the whole. Elsewhere Aglietta (1979, 171) writes that "the reproduction of class society is a single totality in which superstructural forms can exert a considerable influence on the law of capital accumulation". The basic problematic at work in Aglietta's thought is the economic reproduction of class society as a single totality supported by superstructural forms.

In a recent article, David Kotz (1990, 26) argues that Regulation Theory is too structural, while the Social Structure of Accumulation School places a one-sided emphasis on class struggle. The solution that Kotz suggests is a simple synthesis between the two. But this does not really get at the reasons behind the economism and voluntarism that

Kotz sees. I want to suggest that Regulation Theory's economism stems from a failure to achieve adequate separation and clear articulation between abstract value theory, mid-range theory, and historical analysis and that the Social Structure of Accumulation School's "voluntarism" perhaps stems from the abandonment of high-range theory and a collapsing together of mid-range and low-range theory in such a way that the emphasis falls on class subjects at the level of historical analysis. What I am advocating is not a little more structure for the SSA approach and a little less structure for the Regulation approach, but a paradigm shift in which "structure" and "agency" are unpacked across distinct levels of abstraction and distinct regions of social life.

Aglietta's historical analysis is populated with regimes of accumulation and modes of regulation that either stabilize modes of growth under particular regimes of accumulation or fail to do so (i.e. they are either functional or dysfunctional). Capital is reduced to being simply the wage-relation, and even though we are told that the wage-relation is structurally complex, it is a kind of essence for the analysis of history. The decisive historical change, for Aglietta, is the move from intensive to extensive accumulation. This move integrates department two with department one such that mass consumption develops parallel to mass production. A new mode of consumption develops, which by integrating department two into the motion of value, expresses "the complete realization of the wage relation" (1979, 81). Intensive accumulation "creates a new mode of life for the wage-earning class by establishing a logic that operates on the totality of time and space occupied or traversed by its individuals in daily life" (1979, 71). This quotation from Aglietta suggests that historical change occurs because of a logic embedded in the wage-relation that shapes the totality of social life. Such a perspective must produce an economic reductionist form of historical analysis.

11.5 AN ALTERNATIVE

I believe that some of the fundamental weaknesses in Aglietta's formulation of Regulation Theory flow from of his Althusserian assumptions.[12] In particular it is Althusser's conception of "mode of production" that leads him astray. According to this conception the economic is a relatively autonomous set of practices "overdetermined" by the political and ideological. But this is not at all the way Marx theorizes the economic in *Capital.* Marx defines capital as self-expanding value (M – C – M′), and the "self" here is of crucial importance. It means that capital is conceived of as expanding itself by means of the operation of its own commodity-economic logic without the reliance on any extra-economic force whatsoever. In other words, in order for the law of value to be clearly theorized, we must assume a total commodification of social relations such that all inputs and outputs of production are commodified and persons are reduced to being simply bearers of economic categories. At the same time, this total reification of pure capitalism makes capital a totally unique theoretical object differing from all other theoretical objects of social life in its reified characteristics. It is only because of reification that the economic can absorb the social and can achieve a certain automaticity that enables us to theorize an inner logic. Because the political and the ideological lack the self-reifying force of capital, they do not have an inner logic, and must therefore be largely theorized at more concrete levels of analysis. It follows that to mix particularistic considerations of extra-economic force directly into a theory of capital's logic in the abstract and in general is to completely undermine the logical coherence of the theory. In short, in the theory of capital's inner logic, the economic is not overdetermined by the political and ideological, rather the social as a whole is absorbed into the automaticity of the economic.

Total reification implies total objectification: it is the motion of objects that controls the motion of persons as the self-organizing powers of persons become totally absorbed into the self-organizing powers of capital. Social life is theorized entirely from the point of view of capital and its self-expanding logic. But if capital is ontologically unique because of its self-reifying character, it is necessary to be cautious in the manner we connect an abstract logic to more concrete levels of analysis and in the manner we connect the economic with other regions of social life. It is obvious that at the level of history society is never totally reified, and hence capital's impact on social change is only partial. It follows that after clarifying the inner logic of capital, the next task of Marxian political economy is to develop complex mediations connecting the abstract and concrete and the economic with the non-economic.

For this I would propose a distinct level of mid-range theory with the following properties:[13]

(1) The logical necessity and automaticity associated with the inner logic of capital is out of place at the level of mid-range theory. The accumulation of capital at this level of analysis is instead associated with the sort of necessity that is contingent upon a stage-specific way of organizing profit-making.
(2) The automaticity associated with the law of value does not apply to stage theory. Hence, if at all, value categories should be used with extreme caution. In general value categories are converted into abstract types of institutional structures at the level of mid-range theory.
(3) Since reification is less than total at this level, the accumulation of capital includes political, legal, and ideological practices as part of its conditions of existence.
(4) The aim of mid-range theory is to outline the most typical structures of capital accumulation including political, legal, and ideological dimensions characteristic of a stage of capitalist development.
(5) To this end mid-range theory should concern itself with the degree to which elements of socio-economic life are commodified, how this degree is maintained, and how the various commodified and quasi-commodified elements typically interact in stage-specific ways to form a mode of capital accumulation.
(6) Given the centrality of the exploitation of labour to capital accumulation, focus on the stage-specific characteristics of the commodification of labour-power is crucial.
(7) It is important to separate questions of historical change from a mid-range theory of abstract structural types. Failure to do so will inevitably lead in the direction of converting history into little more than a function of a logic of abstract structures. Hence, mid-range theory describes types of structural relations, but does not attempt to explain historical change directly.
(8) Mid-range theory is neither a deduction from the theory of capital's logic nor is it an induction from historical analysis, rather it is a distinct level of analysis informed by both. It is not a theory of global capitalism nor of particular societies, instead it constructs a theory of capital accumulation that portrays the *modus operandi* of capital that are most typical and characteristic of a world-historical stage of capitalist development.

My primary emphasis has been upon the weaknesses of Regulation Theory that stem from its refusal to inform its mid-range theory by a rigourous and autonomous theorization of capital's inner logic.[14] Regulation theorists have tended either to neglect value theory or to convert it into a voluntaristic formal model, and in this latter case they have failed to adequately problematize the relation between levels of analysis. An equally important source of shortcomings is their tendency not to adequately problematize the relation between mid-range theory and historical analysis. Historical change is enormously complex, involving ever-shifting patterns of power at all levels of social life from local and small sets of social relations to global and large sets of social relations. It is always very difficult to ascertain which changes are most important in affecting particular historical outcomes. Furthermore, at the level of historical analysis it is probably a mistake to consider society a totality in the sense of being a unity, even a complexity unified by the structure in dominance. Social life is simply too heterogeneous, open-ended, and unstable to be encapsulated as a unified totality.

From the point of view of historical analysis, the theory of capital's inner logic can clarify what we mean by "capital", and mid-range theory can clarify characteristic stage-specific types of capital accumulation, but both of these more abstract levels of theory can only inform an autonomous historical analysis which must consider capital accumulation as only one force of historical change among others. In order to avoid economic reductionism, it is essential that we approach historical outcomes with an open mind about capital's role as one social force amongst other social forces. Because of capital's reifying force, all things being equal, it tends to prevail over other social forces, but in particular circumstances, when things are seldom equal, other social forces may be more decisive.

The approach that I am advocating gives historical analysis a great deal more territory than is usual in the social sciences. The reason is that in my view mid-range theories as usually constituted in either Marxian or non-Marxian social science nearly always reduce history to being little more than a function of structural mid-range theory. I believe this is a mistake because the complexity of process, change, and agency in history is violated by such approaches. In opposition I am advocating an autonomous historical analysis that is informed by, but not determined by the two more abstract levels of analysis. Instead of having one rather necessitarian logic formed in the homogeneous space of value reaching down across levels of abstraction to determine the concrete, I am suggesting three autonomous logics that inform each other best by keeping their logics to themselves.[15]

If the economic in capitalist societies is best theorized at three relatively autonomous levels of analysis, it follows that there is no general solution to the structure/agency problem. Instead structure and agency will be theorized differently at each distinct level of analysis. Because of the total reification of pure capitalism, persons are in the end reduced by capital's logic to being mere "bearers" or "personifications" of economic categories. The deep structure of capital is pure structure in the sense that persons are reduced in the end to being mere instruments of the structural imperatives of self-expanding value. At the level of stage theory, we look at institutions as abstract types or paradigms, and this would include abstract types of agency. Agency enters in, but it is a highly structured agency. At the level of historical analysis, agency comes to the fore, as all institutions are continually negotiated and renogiated through the ebb and flow of domination and resistance, of empowerment and disempowerment. But at this level it is usually collective agency that interests us, and it is always more or less decentered and complex. Furthermore, "structure" is possibly a misleading concept since what exists is human institutions in process, and "structure" connotes something that is relatively static and without agency. Of course, some of these historical institutions are more enduring, hegemonic, and persistent than others, but they are all agencied. In short, what I am proposing is conceiving of the relation between structure and agency differently depending upon the level of analysis. In this way we can avoid trying to combine them within the same level of analysis where structure tends to be imperialistic over agency as in Regulation Theory or where agency tends to undermine structure, as Kotz claims is characteristic of the Social Structure of Accumulation Approach.

11.6 CONCLUSIONS

I have argued that Aglietta's effort to theorize the economic under capitalism fails to connect the abstract and concrete or the economic and non-economic in ways that can avoid economic reductionism. His value theory is a formal model based on the "homogeneous space of value". The relation between it and mid-range theory is not sufficiently problematized. The result is that the economic at the level of mid-range theory comes to be overly reified with the political and ideological being reduced to add ons. With his conception of homogeneous value space, Aglietta tends to make the homogeneous imperialistic over the non-homogeneous. His emphasis on the wage-relation leads to a one-sidedly productivist conception of capital. His particular appropriation of Marx's Reproduction Schema both introduces an element of classical general equilibrium theory into Marx, and imposes the resulting mechanicalness directly on the social and historical. As a result politics and ideology fall by the wayside or tend to be reduced to a simple function of the economic. Despite many statements sprinkled through Aglietta's book suggesting the primacy of class struggle, these must be viewed as empty rhetoric in the face of the laws of accumulation and competition that reduce history to being a mere function of the inner logic of the capitalist mode of production.

Realizing at least some of the difficulties associated with Aglietta's appropriation of *Capital*, some regulationists have advocated ignoring *Capital* and value categories altogether. Indeed, Aglietta himself seems to have moved in this direction in his more recent work. I believe this is the wrong approach because *Capital* offers us too much to be ignored. Instead, as I have suggested, with a more careful analysis of reification and its relation to abstraction, we can use the concept of a purely capitalist society to inform our more concrete levels of analysis without turning the concrete into a function of a necessitarian abstract economic logic. A central plank of the approach that I am advocating is that it is possible to theorize capital's "inner essence" without falling into the traps of economic reductionism, when autonomous levels of analysis are employed.[16] In other words, a theory of capital's inner logic or "essence" need not be applied directly to more concrete levels of analysis.

While this paper may be viewed as critical of Regulation Theory, I want to end on a more conciliatory note. The five contributions that I listed in the introduction are important contributions. For example, it is clear that I agree with Regulation Theory that levels of analysis are important to Marxian political economy. My critique is that they have not sufficiently problematized and analyzed this crucial issue. My aim is not to dismiss Regulation Theory, but to critically absorb their contributions in the process of proposing a stronger and more effective approach to Marxian political economy.

Notes and References

1. The main influence comes from the work of Uno and Sekine listed under "references", but other Japanese scholars have also influenced me. In particular I would mention Professors Mawatari, Watanabe, Itoh, Nagatani, Yamamoto, and H. Ouchi.
2. For a fuller discussion of the peculiarities of capital as an object of knowledge see Albritton (1991, 1992, 1993b).

3. Marx either uses the term "pure" or analogues in referring to capital's immanent laws too many times to list all the references – a few follow (1976: 203, 260, 261, 793; 1967: 25; 1971: 175, 268 281, 624, 885)
4. For a more extended discussion of the relation between the logical and historical see Albritton (1986).
5. Some Marxists have abandoned value theory because of difficulties with Marx's formulation of the transformation problem. But the fact of the matter is that there is no theory of price determination without technical difficulties, and there is no reason why such a theory at least as strong as competing theories cannot be developed within the framework of theorizing a purely capitalist society based on a profit system rooted in the exploitation of labour. In other words, one can accept the general framework of Marx's *Capital* and work out the difficulties of technical economics within that framework including the relationship between value and price. (See Sekine, 1984, 1986.)
6. The idea of "market mentality" comes from Polanyi (1944).
7. For a thorough discussion of types of labour not valued by capital see Waring (1988).
8. Brenner and Glick (1991) also make this point, however, their critique of Regulation Theory is almost entirely negative and empiricist, offering no alternative when it comes to the decisive questions about precisely how Marx's *Capital* is to be interpreted and utilized.
9. For a very clear presentation of Marx's Reproduction Schema see Sekine (1984). Also see Rosdolsky (1977).
10. "There were 3,653 recorded mergers between 1898 and 1902, twenty-five times the total number in the preceding three years and six times the number in the succeeding five years" (Gordon, Edwards, Reich, 1982, 107).
11. I believe Jessop has misinterpreted the "Introduction" to Marx's *Grundrisse*, where clearly Marx is referring to the concrete-in-thought and not the real concrete as a "synthesis of multiple determinations".
12. For a discussion of the connection between Althusserianism and Regulation Theory see the article by A. Lipietz in E. Kaplan and M. Sprinker (1993).
13. This is a very brief characterization of what I call "stage theory". For an initial effort at theorizing this complex mid-range theory see Albritton (1991).
14. For an example of the stress I place on the autonomy of historical analysis see Albritton (1993a).
15. See Albritton (1991; 1992) for elaboration of the separate "logics" associated with different levels of analysis.
16. For more discussion of how economic "essentialism" can be completely separated from economic reductionism see Albritton (1993b).

Bibliography

Aglietta, M. (1979) *A Theory of Capitalist Regulation*, London: New Left Books.

Albritton, R. (1984) "The Dialectic of Capital: A Japanese Contribution", *Capital and Class*, Vol. 22, Spring.

Albritton, R. (1986) *A Japanese Reconstruction of Marxist Theory*, London: Macmillan.

Albritton, R. (1991) *A Japanese Approach to Stages of Capitalist Development*, London: Macmillan.

Albritton, R. (1992) "Levels of Analysis in Marxian Political Economy: An Unoist Approach", *Radical Philosophy*, No. 60.

Albritton, R. (1993a) "Theorizing the Economic: Weber, Marx and Uno", unpublished manuscript.

Albritton, R. (1993b) "Did Agrarian Capitalism Exist", *The Journal of Peasant Studies*, Vol. 20, No. 3, April, '93.

Albritton, R. (1993c) "Marxian Political Economy for an Age of Postmodern Excess", *Rethinking Marxism*, Vol.6, No.1, Spring '93.

Althusser, L. and Balibar, E. (1970) *Reading Capital*, London: New Left Books.

Amariglio, J. and A. Callari (1989) "Marxian Value Theory and the Problem of the Subject: The Role of Commodity Fetishism", *Rethinking Marxism*, 2:3, pp. 31–60.

Aozai, T. (1978) "Shihon-Ron to Puran-Mondai" (Marx's *Capital* and the Plan Problem), *Keizai-Gaku Hihan*, No. 4, Tokyo: Shakai-Hyoron-Sha.

Baba, H. (1972) "Kyoko-Ron ni okeru Bumonkan-Fukinkou" (On Inter-sectoral Disequilibrium in Crisis Theory), Tohara, Shiro, *Kyoko-Ron*, Tokyo: Chikuma-Shobo.

Baba, H. (1973[1962]) "Kahei to Kyoko" (Money and Crises), *Sekai-Keizai: Kijiku to Shuhen*, Tokyo: Tokyo-Daigaku-Shuppankai.

Baudrillard, J. (1975) *The Mirror of Production*, St. Louis: Telos Press.

Baudrillard, J. (1981) *For A Critique of the Political Economy of the Sign*, St. Louis: Telos Press.

Belsey, C. (1980) *Critical Practice*, New York: Methuen.

Benhabib, S. and Cornell, D., eds. (1987) *Feminism as Critique*, Minneapolis: University of Minnesota Press.

Berg, M. (1985) *The Age of Manufactures: 1700–1820*, New Jersey: Barnes & Noble.

Bertramsen, R., *et al.* (1991) *State, Economy and Society*, London: Unwin Hyman.

Böhm-Bawerk, E. (1949) *Karl Marx and the Close of His System*, London: Merlin Press.

Bossy, J. (1985) *Christianity in the West*, Oxford: Oxford University Press.

Boyer, R. (1990) *The Regulation School: A Critical Introduction*, New York: Columbia University Press.

Braudel, F. (1972, 1973) *The Mediterranean and the Mediterranean World in the Age of Philip II*, Volumes I & II, Harper: New York.
Braudel, F. (1988, 1990) *The Identity of France*, Volumes I & II, London: Collins.
Brenner, R. and M. Glick (1991) "The Regulation Approach: Theory and History", *New Left Review*, No. 188.
Brewer, J., N. McKendrick, and J. H. Plumb (1982) *Birth of a Consumer Society: The Commercialization of 18th Century England*, London: Europa Press.
Brody, D. (1960) *Steelworkers In America*, New York: Harper & Row.
Burnett, J. (1989) *Plenty and Want*, London: Routledge.
Butler, J (1990) *Gender Trouble*, New York: Routledge.
Chayanov, A. V. (1966) *On the Theory of Peasant Economy*, Homewood, Illinois: American Economic Association.
Clarke, S. (1988) "Overaccumulation, Class Struggle and the Regulation Approach", *Capital and Class*, No. 36.
Cornell, D. (1991) *Beyond Accommodation*, New York: Routledge.
Cronon, W. (1991) *Nature's Metropolis*, New York: Norton.
Crouzet, F. (1982) *The Victorian Economy*, New York: Columbia University Press.
Daly, G. (1991) "The Discursive Construction of Economic Space: Logics of Organization and Disorganization", *Economy and Society*, 20:1, pp. 79–102.
Darby, H. C., ed. (1976) *A New Historical Geography of England After 1600*, Cambridge: Cambridge University Press.
Davidoff, L. and Hall, C. (1987) *Family Fortunes*, London: Hutchinson.
Davis, M. (1978) "Fordism in Crisis", *Review* II, 2.
Debreu, G. (1959) *Theory of Value*, New York: Wiley.
DeMartino, G. (1992) *Modern Macroeconomic Theories of Cycles and Crisis: A Methodological Critique*, Ph.D. dissertation, University of Massachusetts Dept. of Economics.
DeMartino, G. (1993) "The Necessity/Contingency Dualism in Marxian Crisis Theory: The Case of Long-Wave Theory", Paper presented at the URPE conference.
Domar, E. D. (1957) *Essays in the Theory of Economic Growth*, Oxford: Oxford University Press.
Duncan, C. (1983) "Under the Cloud of Capital: History vs. Theory", *Science and Society*, 47.
Duncan, C. (1986) "On Rapid Industrialization and Collectivization: An Essay in Historiographic Retrieval and Criticism" *Studies in Political Economy*, 21.
Duncan, C. (1991) "On Identifying a Sound Environmental Ethic in History: Prolegomena to Any Future Environmental History", *Environmental History Review*, 15.
Duncan, C. (1992) "Legal Protection for the Soil of England: The Spurious Context of Nineteenth Century 'Progress'", *Agricultural History*, 66.
Duncan, C. (forthcoming) *The Centrality of Agriculture: Between Humankind and the Rest of Nature*, Kingston and Montreal: McGill-Queen's University Press.
Duncan, C. and M. Maruyama (1990) "The Japanese Counterpart to Karl Polanyi: The Power and Limitations of Kozo Uno's Perspective", *The Life and Work of Karl Polanyi*, ed. K. Polany-Levitt, Montreal: Black Rose.

Dunne, P. (1977) "Hegel's Doctrine of Quality in Reference to the Theory of the Commodity-Form in the Dialectic of Capital", unpublished manuscript.
Ekins, P., ed. (1986) *The Living Economy*, London: Routledge & Kegan Paul.
Ewen, S. (1976) *Captains of Consciousness*, New York: McGraw-Hill.
Featherstone, M. (1991) *Consumer Culture and Postmodernism*, London: Sage Press.
Fraad, H., Resnick, S. and Wolff, R. (1989) "For Every Knight in Shining Armor, There's a Castle Waiting To Be Cleaned: A Marxist-Feminist Analysis of the Household", *Rethinking Marxism*, Vol. 2, No. 4.
Fraser, N. and L. J. Nicholson (1990) "Social Criticism Without Philosophy: An Encounter Between Feminism and Postmodernism", *Feminism/Postmodernism*, ed. L. J. Nicholson, New York: Routledge, pp. 19–38.
Furihata, S. (1965) *Shihonron Taikei no Kenkyu* (Studies in the System of *Capital*), Tokyo: Aoki-shoten.
Furukawa, T. (1959) "Uno-Kyoju-Kyoko-Ron no Mondaiten" (Some Problems with Professor Uno's Crisis Theory), *Keizai-Hyoron*, April.
Gibson-Graham, J. K. (1993) "Waiting for the Revolution, or How to Smash Capitalism while Working at Home in Your Spare Time", *Rethinking Marxism* 6:2, pp. 10–24.
Gordon, D., R. Edwards, and M. Reich (1982) *Segmented Work, Divided Workers*, Cambridge: Cambridge University Press.
Grossman, H. (1932[1929]) *Shihon no Chikuseki narabini Houkai no Riron* (Das Akkumulations-und Zusammenbruchsgesetz des kapitalistishen Systems), trans. by H. Arisawa and K. Moriya, Tokyo: Kaizo-Sha.
Harris, F. (1992) *A Passion for Government*, Oxford: Oxford University Press.
Haruta, M. (1974) ["Demonstration of the Law of Value", *Seminar in Economics: 1 Marxian Economics*], ed. K. Suzuki, Tokyo: Nihonhyoron-sha.
Harvey, D. (1989) *The Condition of Postmodernity*, Oxford: Basil Blackwell.
Hayek, F. A. (1976) *Denationalization of Money*, London: The Institute of Economic Affairs.
Heaton, H. (1965) *The Yorkshire Woollen and Worsted Industries*, Oxford: The Clarendon Press.
Hegel, G. W. F. (1969) *Logic*, trans. by A.V. Miller, London: Allen & Unwin.
Hegel, G. W. F. (1975) *Logic*, trans. by W. Wallace, Oxford University Press.
Hidaka, H. (1983) *Keizai Genron* (Principles of Political Economy), Tokyo: Yuhikaku.
Hidaka, H. (1987) *Shihon-Chikuseki to Keiki-Junkan* (Capital Accumulation and Business Cycles), Tokyo: Hoseidaigaku-Shuppankai.
Hilton, B. (1988) *The Age of Atonement: The Influence of Evangelists on Social and Economic Thought, 1785–1865*, Oxford: Oxford University Press.
Hirsch, F. (1976) *Social Limits to Growth*, Cambridge: Harvard University Press.
Hirschman, W. O. (1977) *The Passions and the Interests: Political Arguments for Capitalism Before its Triumph*, Princeton: Princeton University Press.
Hirst, P. and J. Zeitlin (1991) "Flexible Specialization Versus Post-Fordism: Theory, Evidence and Policy Implications", *Economy and Society*, Vol. 20, No. 1.
Hoshino, T. (1977) "Shihonkajo to Kyoko" (The Excess of Capital and Crises), *Keizai-Gaku*, Tohoku University, Vol. 39, No. 3.

Hoshino, T. (1980) "Shogyo-Shin'yo Dotai no Kihon-Genri" (The Basic Theory of Dynamics of Commercial Credit), *Keizai-Gaku*, Tohoku University, Vol. 42, No. 2.
Hoshino, T. (1987) "Ginko-Shin'yo Dotai no Kihon-Genri" (The Basic Theory of Dynamics of Bank Credit), *Keizai-Gaku*, Tohoku University, Vol. 48, No. 4.
Hoshino, T. (1989) "Keiki to Kyoko" (Business Cycles and Crises), ed. by S. Mawatari, *Economics at the Present*, Kyoto: Shohwa-Do.
Hoskins, W. G. (1970) *The Making of the English Landscape*, Harmondsworth: Penguin.
Ishibashi, S. (1992) *Sihon To Rijun* (Capital and Profit), Tokyo: Zeimu-keiri-kyokai.
Itoh, M. (1973) *Shin'yo to Kyoko* (Credit and Crises), Tokyo: Tokyo Daigaku-Shuppankai.
Itoh, M. (1980) *Value and Crisis*, New York: Monthly Review Press.
Itoh, M. (1988) *The Basic Theory of Capitalism*, London: Macmillan.
Iwata, H. (1967) *Marukusu Keizaigaku* (Marxian Economics), Tokyo: Morita-shoten.
Jenkins, R. *The Road to Alto*, London: Pluto.
Jessop, B. (1988) "Regulation Theory, Post Fordism and the State", *Capital and Class*, No. 34.
Jessop, B. (1990) "Regulation Theories in Retrospect and Prospect", *Economy and Society*, Vol. 19, No. 2.
Kamakura, T. (1970) *Shihonron Taikei no Hoho* (Method of the System of Capital), Tokyo: Nihonhyoron-sya.
Kassiola, J. (1990) *The Death of Industrial Civilization*, Albany: SUNY Press.
Kawai, I. (1957) "Jitsugen-Ron naki Kyoko-Ron" (The Crisis Theory without Realization Theory), *Shisou*, November.
Kellner, D. and Best, S. (1991) *Postmodern Theory*, New York: Guilford.
Kobayashi, Y. (1976) "Kyoko no Hitsuzensei" (The Necessity of Crises), ed. Keizairiron-Gakkai, *Gendaishihonshugi to Kyoko*, Tokyo: Aoki-Shoten.
Kobayashi, Y. (1977) Kachiron to Tenkei-Mondai Ronso, (The Theory of Value and the Debate on the Transformation Problem), Tokyo: Ochanomizu-shobo.
Kobayashi, Y. (1979) "Kyoko no Hitsuzensei to Fukinkou" (The Logical Necessity of Crises and Intersectoral Disequilibrium), *Tsukuba-Daigaku: Keizaigaku-Ronshu*, No. 3.
Kobayashi, Y. (1981) "Rodo-Kachisetsu no Ronsho Mondai (2)" ("The Problem of Deonstration of the Labour Theory of Value, 2'), *Keizaigaku-ronshu*, Vol. 7, Tsukuba University.
Kohr, L. (1978) *The Over-Developed Nations*, New York: Schocken.
Kotz, D. (1990) "A Comparative Analysis of the Theory of Regulation and the Social Structure of Accumulation Theory", *Science and Society*, Vol. 54, No. 1.
Kroker, A. and M. Kroker (1991) *Ideology and Power*, Montreal: New World Perspectives.
Kurita, Y. (1992) *Kyoso to Keikijunkan* (Competition and Business Cycles), Tokyo: Gakubunsha.

Kurth, J.R. (1979) "The Political Consequences of the Product Cycle: Industrial History and Political Outcomes", *International Organization* 33:1, Winter.

Laclau, E. (1990) *New Reflections on the Revolution of Our Time*, London: Verso.

Laclau, E. and C. Mouffe (1985) *Hegemony and Socialist Strategy*, London: Verso.

Lipietz, A. (1993) "From Althusserianism to Regulation Theory", *The Althusserian Legacy*, eds. E.A. Kaplan and M. Sprinker, London: Verso.

MacLean, B. (1981) "Kozo Uno's Principles of Political Economy", *Science and Society*.

Mandel, E. (1978) *Late Capitalism*, London: Verso.

Mandel, E. (1980) *Long Waves of Capitalist Development*, Cambridge: Cambridge University Press.

Maruyama, M. (1988) "Local Currency as a Convivial Tool", *The Meiji Gakuin Review: International and Regional Studies* no. 3, Tokyo: Meiji Gakuin University.

Maruyama, M. (forthcoming) "Hansatsu: Local Currencies in Pre-Industrial Japan" eds. C. Duncan, and D. Tandy, *From Political Economy to Anthropology*, Montreal: Black Rose Books.

Marx, K. (1967) *Capital*, Vol. II, Moscow: Progress.

Marx, K. (1971a) *Capital*, Vol. I, Moscow: Progress.

Marx, K. (1971b) *Capital*, Vol. III, Moscow: Progress.

Marx, K. (1973) *Grundrisse*, London: New Left Review.

Marx, K. (1974) *Theorien uber den Mehrwert*, Vol. II, Berlin: Diez Verlag.

Marx, K. (1976) *Capital*, Vol. I, London: New Left Books.

Matsuda, M. (1990) "Kyoko to Shin'yo-Seido" (Crises and the Credit System), *Keizairiron-Gakkai-Nenpo*, No. 27.

Mawatari, S. (1970) "Shihonka to Rodosha no Kankei to Shihonka to Shihonka no Kankei" ("The Capitalist-Worker Relation and the Inter-Capitalist Relation"), in *Shihonron to Teikokushugiron* (*Capital* and the Theory of Imperialism) , Vol. 1, ed. T. Takeda, *et al.*, Tokyo: University of Tokyo Press.

Mawatari, S. (1973a) "Shihon-Ippan to Kyoko-Ron" (Capital in General and the Theory of Economic Crises), *Keizai Shirin*, Hosei University, Vol. 42, No. 3/4.

Mawatari, S. (1973b) "Keiki-Junkan-Katei: 1830–Nendai niokeru" (The Process of Business Cycles in 1830's), ed. K. Suzuki, *Kyoukoshi-Kenkyu* (History of Business Cycles), Tokyo: Nihon-Hyoron-Sha.

Mawatari, S. (1973c) "Shihon no Saiseisan-Katei I/II" (The Reproduction Process of Capital), *Daigakukoza-Keizaigaku: Shihon-Ron to Gendai*, Tokyo: Nippon-Hosokyokai.

Mawatari, S. (1977) "Sutagufurehshon" (Stagflation), Rodo-Undo-Kenkyusha-Shudan, ed., *Kaikyuteki-Rodo-Undo eno Mosaku*, Tokyo: Nihon-Hyoron-Sha.

Mawatari, S. (1979a) "Kyoko-Ron to Gendai-Shihonshugi" (Economic Crisis Theory and Contemporary Capitalism), *Keizai-Hyoron*.

Mawatari, S. (1979b) "Kachiron Ronso no Gen-Chiten" (The Present State of the Debate on the Theory of Value), *Keizai-Hyoron*, Dec.

Mawatari, S. (1985) "The Uno School: A Marxian Approach in Japan", *History of Political Economy Journal*, Vol. 17, No. 3.
McCracken, G. (1990) *Culture and Consumption*, Bloomington, Indiana: Indiana University Press.
Miwa, H. (1987) "Kachi-Hosoku no Ronsho" (The Demonstration of the Law of Value), *Keizaigaku-Ronkyu*, Vol. 7, Tsukuba University.
Morishima, M. (1973) *Marx's Economics*, Cambridge: Cambridge University Press.
Mui, H., *et al.* (1989) *Shops and Shopkeeping in Eighteenth Century England*, Montreal and Kingston: McGill-Queen's University Press.
Nagatani, K. (1981) *Kachiron no Shin-Chihei* (A New Horizon of the Theory of Value), Tokyo: Yuhikaku.
Neale, W. C. (1976) *Monies in Societies*, San Francisco: Chandler & Sharp.
Neeson, J. (1993) *Commoners: Common Right, Enclosure and Social Change in England, 1700–1820*, New York: Cambridge University Press.
Norton, B. (1988) "Epochs and Essences: A Review of Marxist Long-Wave and Stagnation Theories" *Cambridge Journal of Economics*, No. 12, pp. 203–224.
Obata, M. (1984a, 1984b), "Joyo no Gainen to Kachi-Hosoku no Ronsho" (The Concept of Surplus and the Demonstration of Value Theory), *Keizaigaku-ronshu*, 50–1, 50–2, Tokyo University.
Offer, A. (1989) *The First World War*, Oxford: Oxford University Press.
Oh'uchi, H. (1966) *Keiki to Kyoko* (Business Cycles and Crises), Tokyo: Kinokuniya-Shoten.
Oh'uchi, H. (1977) "Jitsugen Kyoko-Ron no Saihan ni-yosete" (On a New Version of Realization Crisis Theory), *Keizaigakuhihan*, No. 2.
Oh'uchi, T. (1982) *Keizai-Genron: Ge* (Principles of Political Economy, Part II), Tokyo: Tokyo-Daigaku-Shuppankai.
Oh'uchi, T. (1983) *Kokka-Dokusen-Shihonshugi: Hatan no Kozo* (The State Monopolistic Capitalism – The Structure of its Failure), Tokyo: Ochanomizu-Shobo.
Okishio, N. (1976) *Chikuseki-Ron (Dai 2–Han)* (Theory of Capital Accumulation), Tokyo: Chikuma-Shobo.
Okishio, N. (1977) *Marukusu Keizaigaku* (Marxian Economics), Tokyo: Chikuma-shobo.
Okishio, N. and M. Itoh (1987) *Keizai-Riron to Gendai-Shihon-Shugi*, Tokyo: Iwanami-Shoten.
Okuyama, T. (1989) "Kakaku, Hiyou to Rijun" (Prices, Costs and Profits), ed. S. Mawatari, *Keizaigaku no Genzai*, Kyoto: Showa-Do.
Polanyi, K. (1944) *The Great Transformation*, Boston: Beacon Press.
Polanyi, K. (1968) *Primitive, Archaic, and Modern Economies*, ed. G. Dalton, Boston: Beacon Press.
Polanyi, K. (1977) *The Livelihood of Man*, New York: Academic Press.
Resnick, S. and R. Wolff (1987) *Knowledge and Class*, Chicago: University of Chicago Press.
Resnick, S. and R. Wolff (1988) "Marxian Theory and the Rhetorics of Economics" *The Consequences of Economic Rhetoric*, eds. A. Klamer, *et al.*, Cambridge and New York: Cambridge University Press, pp. 47–63.

Rorty, R. (1985) "Solidarity or Objectivity" *Post-Analytic Philosophy*, eds. J. Rajchman and C. West, New York: Columbia University Press, pp. 3–19.
Rosdolsky, R. (1977) *The Making of Marx's "Capital"*, London: Pluto Press.
Sahlins, M. (1976) *Culture and Practical Reason*, Chicago: University of Chicago Press.
Sakaguchi, M. (1960) "Shihonkosei-Kohdoka to 'Rijunritsu-Keikou-Teika'" (Rise in the Organic Composition of Capital and "The Tendency of the Rate of the Profit to Fall"), ed. K. Suzuki, *Rijun-Ron-Kenkyu*, Tokyo: Tokyo-Daigaku-Shuppankai.
Sakurai, T. (1968) *Seisan-Kakaku no Riron* (The Theory of the Prices of Production), Tokyo: University of Tokyo Press.
Samuelson, P. A. (1957) "Wages and Interest: A Modern Dissection of Marxist Economic Models", *The American Economic Review* XLVII.
Schurtz, H. (1898) *Grundriss einer Entstehungsgeschichte des Geldes*, Weimer: Verlag von Emil Felber.
Sekine, T. (1975) "Uno-Riron: A Japanese Contribution to Political Economy", *Journal of Economic Literature*, Vol. XIII.
Sekine, T. (1980) "The Necessity of the Law of Value", *Science and Society* XLIV.
Sekine, T. (1982) "Economic Theory and Capitalism", *York Studies in Political Economy*, I.
Sekine, T. (1983) "The Law of Market Value", *Science and Society*, XLVI.
Sekine, T. (1984) *The Dialectic of Capital*, Vol. 1, Tokyo: Yushindo Press (reissued in 1986 by Toshindo Press).
Sekine, T. (1985a) "The Pricing of Commodities", *York Studies in Political Economy*, IV.
Sekine, T. (1985b) "The Transformation Problem, Qualitative and Quantitative", *York Studies in Political Economy*, VI, pp. 60–96.
Sekine, T. (1986) *The Dialectic of Capital*, Vol. 2, Tokyo: Toshindo Press.
Sekine, T. (1993) "Introduction to the Method of the Dialectic of Capital and the Theory of Value", Paper presented at the Uno Conference York University.
Seton, F. (1957) "The Transformation Problem", *The Review of Economic Studies* XX.
Smith, D. (1989) "Feminist Reflections of Political Economy", *Studies in Political Economy*, No. 30, Autumn.
Stigler, G. J. (1965) *Essays in the History of Economics*, Chicago: University of Chicago Press.
Sugiura, K. (1975) "Kyoko no Kiso-Riron" (The Basic Theory of Crises), *Tokyo-Daigaku: Shakaikagaku-Kiyou*, No. 24.
Suzuki, K., ed. (1960) *Keizaigaku Genriron*, Vol. I (Principles of Political Economy, vol. I) Tokyo: Tokyo Daigaku Shuppankai.
Suzuki, Koh'ichiro, ed. (1962) *Keizaigaku-Genriron: Ge* (Priciples of Political Economy, vol. II), Tokyo: Tokyo Daigaku ShuppanKai.
Suzuki, Kikuo (1966) "Shihon no Zettaiteki-Kajoseisan ni tsuite" (on the Absolute Over-Production of Capital), *Yuibutsu-Shikan*, No. 3.
Sweezy, P. (1942) *The Theory of Capitalist Development*, New York: Monthly Review Press.

Sweezy, P., ed. (1949) "Karl Marx and the Close of His System" by Eugene von Bohm-Bawerk & Bohm-Bawerk's Criticism of Marx" by Rudolph Hilferding, New York: Augustus Kelley.
Sweezy, P. (1967) *Shihonshugi-Hatten no Riron* (The Theory of Capitalist Development), trans. by S. Tsuru, Tokyo: Shin'hyoron.
Takasuka, Y. (1979) *Marukusu-Keizaigaku Kenkyu* (Studies in Marxian Economics), Tokyo: Shin-Hyoron.
Takasuka, Y. (1985) *Marukusu no Kyoso-Kyoko-Kan* (Marx's Views on Competition and Crises), Tokyo: Iwanami-Shoten.
Takei, K. (1985) "Keiki-Junkan no Riron" (The Theory of Business Cycles), eds. K. Takei, I. Okamoto, K. Ishigaki, *Keiki-Junkan no Riron*, Tokyo: Jicho-Sha
Tawney, R.H. (1938) *Religion and the Rise of Capitalism*, Harmondsworth: Penguin.
Therborn, G. (1976) *Science, Class and Society*, London: NLB.
Thompson, E.P. (1968) *The Making of the English Working Class*, Harmondsworth: Penguin.
Thompson, E.P. (1991) *Customs in Common*, New York: The New Press.
Tohara, Shiro (1972) Kyokoron (The Theory of Economic Crises), Tokyo: Chikuma Shobo.
Trigger, B. (1985) *Natives and Newcomers*, Montreal and Kingston: McGill-Queen's.
Tugan-Baranovsky (1972 [1901]) *Eikoku-Kyokoshi-Ron* (Studien zur Theorie und Geshichte der Handelskrisen in England), trans. by S. Kunigo, Tokyo: Perikan-Sha.
Uno, K. (1953) *Kyoko-Ron* (Theory of Economic Crisis), Tokyo: Iwanami-Shoten.
Uno, K. (1959) "Kyoko no Hitsuzensei wa ikani Ronsho sarerubekika" (How Should the Logical Necessity of Crisis be Demonstrated?), *Shisou*, Jan.
Uno, K. (1962a) *Keizaigaku-Hoho-Ron* (Methodology of Economics), Tokyo: Tokyo-Daigaku Shuppan-Kai.
Uno, K. (1962b) "Kachiron no Ronsho ni tsuite" ("On the Problem of the Demonstration of the Theory of Value") in Uno, K. Keizaigaku-Hohoron (Methodology of Political Economy), Tokyo: Daigaku Shuppankai.
Uno, K. (1964) *Keizai Genron* (Princples of Political Economy) Tokyo: Iwanami Shoten.
Uno, K. (1971) *Keizai Seisakuron* (Types of Economic Policy under Capitalism) Tokyo: Kobundo Press.
Uno, K. (1980) *Principles of Political Economy: Theory of a Purely Capitalist Society*, trans. by T. Sekine, Sussex: Harvester Press.
Uno, K. *Types of Economic Policies Under Capitalism*, translation of *Keizaiseisakuron* (under preparation).
Waring, M. (1988) *If Women Counted*, San Francisco: Harper.
Watanabe, A. (1975a, 1975b) "Kachi to Seisan-Kakaku (3), (4)" (Values and Production-Prices 3, 4), *Keizai-Riron*, Vols. 147, 148, Wakayama University.
Weber, M. (1958) *The Protestant Ethic and the Spirit of Capitalism*, New York: Scribner's.
Winfield, R. (1988) *The Just Economy*, London: Routledge.

Woodell, S. R. J., ed. (1985) *The English Landscape*, Oxford: Oxford University Press.

Yamaguchi, S. (1984) *Kin'yu-Kiko no Riron* (The Theory of Credit Mechanism), Tokyo: Tokyo-Daigaku-Shuppankai.

Yamaguchi, S. (1985) *Keizai Genron Kogi* (Lectures on the Principles of Political Economy), Tokyo: University of Tokyo Press.

Yamaguchi, S. (1987) *Kachiron no Shatei* (The Scope of the Theory of Value), Tokyo: University of Tokyo Press.

Yamaguchi, S., T. Sakurai, and M. Itoh eds. (1985) *Kyoko-Ron no Shin-Tenkai* (New Development of Crisis Theory), Tokyo: Shakai-Hyoronsha.

Yamaguchi, S., M. Takumi, and M. Itoh eds. (1979) *Kyoso to Shin'yo* (Competition and Crises), Tokyo: Yuhikaku.

Note: Titles [in brackets] are translated from the Japanese.

Index